Set Theory *and the* Continuum Hypothesis

Paul J. Cohen

with a new introduction by
Martin Davis

Dover Publications, Inc.
Mineola, New York

Copyright

Bibliographical Note

This Dover edition, first published in 2008, is an unabridged republication of the work first published by W. A. Benjamin, Inc., in 1966, and includes a new introduction by Martin Davis. "My Interaction with Kurt Gödel: The man and his work" from *Kurt Gödel and the Foundations of Mathematics: Horizons of Truth*, by Paul J. Cohen, is reprinted courtesy of Cambridge University Press, 2009.

Library of Congress Cataloging-in-Publication Data

Cohen, Paul J., 1934–2007.
Set theory and the continuum hypothesis / Paul J. Cohen. — Dover ed.
p. cm.
"This Dover edition, first published in 2008, is an unabridged republication of the work first published by W. A. Benjamin, Inc., in 1966, and includes a new introduction by Martin Davis."
ISBN-13: 978-0-486-46921-8
ISBN-10: 0-486-46921-2
1. Set theory. 2. Logic, Symbolic and mathematical. Continuum hypothesis. I. Title.

QA248.C614 2008
511.3'22—dc22

2008042847

Manufactured in the United States by LSC Communications
4500058110
www.doverpublications.com

Foreword

Although Paul Cohen's reputation will forever be defined by his Fields Medal winning contributions to logic and set theory, he was equally famous amongst friends and colleagues for his stunning breadth and depth of mathematical knowledge. He could not only give the proof of any great theorem you could name, but in all likelihood it would be one more elegant and beautiful than you had ever seen before. Later when you came back to him frustrated that you could not find that particular method in any textbook, he would explain that was because he had come up with it himself.

For Paul was one of the greatest auto-didacts in twentieth century mathematics; most famously of course in logic and set theory, where in two years he went from knowing almost nothing about the field to solving its most famous problem. Through this process of discovering subjects anew he came up with insightful methods that were legendary in his teaching career.

Unfortunately few of these examples survive in written form, which is why we are so pleased that this book is being reissued. Based on a course that Paul taught at Harvard shortly after his work on the continuum hypothesis, it serves as a small glimpse into an amazing mind that could strip away the technical complications of mathematics leaving its beauty clear to see. Paul used to say that his first love was really music, not math, and we think that any student of mathematics will hear the musicality of this work.

Charles Cohen
October 2008

Introduction to the Dover Edition

The great mathematician David Hilbert was deeply upset. His brilliant student Hermann Weyl had gone over to the dark side, exclaiming: "Brouwer, that is the revolution!" Weyl had joined the rebel Brouwer in proposing to abandon the rigorous foundations of calculus provided by Weierstrass and Dedekind along with Cantor's set theory and transfinite numbers. Speaking in 1922 in language echoing the political and social turmoil of post-war Germany, Hilbert exclaimed:

> What Weyl and Brouwer are doing amounts in essence ... to provide a foundation for mathematics by pitching overboard whatever discomforts them and declaring an embargo ... But this would mean dismembering and mutilating our science, and, should we follow such reformers, we would run the risk of losing a large part of our most valued treasures. Weyl and Brouwer outlaw the general notion of irrational number, of function, even of number-theoretic function, Cantor's [ordinal] numbers of higher number classes, etc. The theorem that among infinitely many natural numbers there is always a least, and even the logical law of the excluded middle, e.g., in the assertion that either there are only finitely many prime numbers or there are infinitely many: these are examples of forbidden theorems and modes of inference. I believe that impotent as [earlier efforts were] to abolish irrational numbers ... no less impotent will their efforts prove today. No! Brouwer's [program] is not as Weyl thinks, the revolution, but only a repetition of an attempted putsch with old methods, that in its day was undertaken with greater verve yet failed utterly. Especially today, when the state power is thoroughly armed and fortified by the work of Frege, Dedekind, and Cantor, these efforts are foredoomed to failure.

As I write more than eighty years later, none of the philosophical issues that excited such passion have really been resolved. Hilbert's own program turned out to be a failure: he had proposed to use mathematical logic to demonstrate the soundness of the mathematics of Cantor and Dedekind by methods that even Brouwer would accept, but Gödel's work effectively blocked this effort. Although a number of mathematicians following in Brouwer's footsteps do advocate similar restrictions, they constitute a very small minority, and it is safe to say that the great bulk of mathematicians working in the mainstream are hardly concerned with such matters. The set-theoretic methods that originally provoked the brouhaha have gradually received general acceptance as they have come to be seen as useful and harmless. It is significant in this respect that the validity of Andrew Wile's 1995 proof of Fermat's Last Theorem has not been questioned despite the heavily set-theoretic framework in which it was presented.

From a pragmatic point of view, mathematical practice today is supported by a foundation consisting of the rules of logical reasoning proposed in 1879 by Gottlob Frege together with the system of axioms of set theory associated with the names Zermelo and Fraenkel. Frege found that to uncover the logical structure of mathematical propositions, one needs not only Boole's propositional connectives:

$\sim$ or $\neg$	not
& or $\wedge$	and
$\vee$	or
$\rightarrow$ or $\supset$	implies
$\leftrightarrow$	if and only if

but also, the quantifiers:

$\exists$	there exists
$\forall$	for all

Making use of this analysis, Frege was able to specify rules for manipulating the symbols so as to yield the effect of the steps of a logical deduction in an ordinary mathematical proof. These rules, often called "first order" logic, can be specified in an number of different ways all of which are equivalent. (A version of these rules will be found on pp. 9-11 of this book.) One way to formulate the rules is in terms of removing and replacing the quantifiers. From this point of view the quantifiers need to be removed because they get in the way of manipulating the propositional connectives. So the rules involve the three moves:

- remove quantifiers
- manipulate propositional connectives
- replace quantifiers

Naturally, the conditions under which these moves can be applied correctly need to be carefully specified. In their everyday work mathematicians in effect are carrying out these moves relying on intuition without the need for the heavy-handed formal apparatus of mathematical logic. But that apparatus comes into its own when it is used to view mathematical activity from "outside" obtaining important theorems about the reach and limitations of mathematical activity.

If Frege's rules of first order logic encapsulate the steps of deduction in a mathematical proof, set theory provides a rich vocabulary of abstract objects that can be used to represent the inventory of objects of ordinary mathematical discourse. Cantor famously explained the notion of a set as

> ...the taking together into a whole of distinct well-defined objects of our intuition or thought.

But from where is one to obtain these "distinct well-defined objects" that are to be gathered into a set? By an ingenious sleight of hand modern set theory sees these objects as being themselves nothing but sets. Axiom 2, in the author's listing of the Zermelo-Fraenkel axioms for set theory, asserts the existence of a set with no members. Axiom 1 guarantees that there is only one such set, the empty set, written $\emptyset$. Applying the various operations on sets that the axioms (listed on pp. 51-53) permit, an enormous menagerie of sets is obtained. From among those, von Neumann suggested identifying the natural numbers $0, 1, 2, 3, \ldots$ with the successive terms of the sequence of sets

$$\emptyset\ ,\ \{\emptyset\}\ ,\ \{\emptyset,\{\emptyset\}\}\ ,\ \{\emptyset,\{\emptyset\},\{\emptyset,\{\emptyset\}\}\ ,\ \ldots$$

where each term of the sequence consists of the sets preceding it in the sequence. We can express this identification by the sequence of equations

$$0=\emptyset\ ,\ 1=\{0\}\ ,\ 2=\{0,1\}\ ,\ 3=\{0,1,2\}\ ,\ \ldots$$

This scheme in which each successive set consists of all of its predecessors can be continued. Thus, after all the natural numbers comes the set

$$\omega=\{0,1,2,3,\ldots\}$$

whose existence is implied by one of the Zermelo-Fraenkel axioms (Axiom 5 in this book). Along these lines, von Neumann's way of dealing with the natural numbers extends in a natural way to provide a way of representing Cantor's transfinite ordinal numbers. As usual, one writes $a \subseteq b$ to mean that every member of a is also a member of b. Following Kuratowski, one can identify the ordered pair $< a, b >$ with the set $\{\{a\},\{a,b\}\}$. This works because the crucial property

$$\text{if } < a,b >=< c,d > \text{ then } a=c \text{ and } b=d$$

is easily proved. Relations and functions can be defined as sets of ordered pairs, and the usual number systems of analysis can likewise be developed.

Thus all the objects that occur in ordinary mathematics can be interpreted as sets. That these sets are obtained from $\emptyset$ without presupposing the existence of any individuals is sometime expressed by calling them "pure" sets. All pure sets are obtainable from $\emptyset$ by indefinitely iterating the *power set* operation, the operation of forming the set of all subsets of a given set:

$$\mathcal{P}(x) = \{y \mid y \subseteq x\}$$

(Axiom 7 states that the power set of any given set exists.) We may think of the power set operation as generating the pure sets in successive layers indexed by Cantor's ordinal numbers, forming what is known as the *cumulative hierarchy*:

$$V_0 = \emptyset; \; V_{\alpha+1} = \mathcal{P}(V_\alpha);$$

$$V_\lambda = \bigcup_{\alpha<\lambda} V_\alpha \quad (\lambda \text{ a limit ordinal})$$

Cantor studied the relation that holds between two sets when there exists a function that provides a one-one correspondence between them. He associated with each set u a *cardinal number* $\overline{\overline{u}}$ with the key property that there is a one-one correspondence between sets u and v if and only if $\overline{\overline{u}} = \overline{\overline{v}}$. He wrote $\aleph_0$ for $\overline{\overline{\omega}}$, and proved that there is no one-one correspondence between ω and $\mathcal{P}(\omega)$, that $\aleph_0 < \overline{\overline{\mathcal{P}(\omega)}}$. Cantor conjectured the following, and tried very hard to prove it without success:

If u is an infinite subset of $\mathcal{P}(\omega)$, then either $\overline{\overline{u}} = \aleph_0$ or $\overline{\overline{u}} = \overline{\overline{\mathcal{P}(\omega)}}$.

This conjecture became known as the *Continuum Hypothesis* (abbreviated CH). When David Hilbert presented his famous list of problems in the year 1900, CH was the first on the list. Yet when Gödel worked on the problem in the 1930s, there still had been no progress towards determining whether or not CH is true. Gödel proved that CH can not be disproved from the Zermelo-Fraenkel axioms, thus if these axioms are consistent, they remain so if CH is added as an additional axiom.

Gödel's proof works with expressions used to define sets written in terms of the logical operations listed above together with the $\in$ of set membership. For example, the expression $\{u \mid u \in x \vee u \in y\}$ in which x and y serve as *parameters* is said to *define* the set $x \cup y$, the union of x and y. Likewise the expression $\{u \mid \exists v(u \in v \wedge v \in x)\}$ defines the set of elements of elements of x, i.e. the union of all the sets that are members of x. For S a given set, we write

$\mathcal{D}(S)$ for the collection of all subsets of S that can be defined in this manner with elements of S being permitted to serve as parameters. Gödel considered a hierarchy of sets like the cumulative hierarchy mentioned above, but with the power set operator $\mathcal{P}$ replaced by $\mathcal{D}$:

$$L_0 = \emptyset; \; L_{\alpha+1} = \mathcal{D}(L_\alpha);$$

$$L_\lambda = \bigcup_{\alpha<\lambda} L_\alpha \quad (\lambda \text{ a limit ordinal})$$

A set S is *constructible*[1] if $S \in L_\alpha$ for some α. Gödel designated by the letter "A" the statement:

Every set is constructible.

He proved that A is consistent with the Zermelo-Fraenkel axioms and that it implies CH. Thus it follows that CH is likewise consistent with these axioms.[2]

Over the years, Gödel expressed varying views concerning the statement A. In his initial announcement of 1938 he said:

> The proposition A added as a new axiom seems to give a natural completion of the axioms of set theory, in so far as it determines the vague notion of an arbitrary infinite set in a definite way.

In a lecture he gave soon afterwards he conjectured that it would turn out that A can not be proved from the Zermelo-Fraenkel axioms. He related the situation that would result to that of Euclid's parallel postulate: just as there is a non-Euclidean geometry in which the parallel postulate is false, so set theory might bifurcate into versions in which A did and did not hold. However, in a famous expository article published in 1947[3] Gödel makes it clear that he now believes A to be false, conjectures that CH will eventually be shown to be independent of the Zermelo-Fraenkel axioms, and predicts that additional more powerful axioms will be found leading to a proof that CH is false.

Paul J. Cohen was born in 1934 of Polish Jewish immigrants. He grew up in Brooklyn and attended Stuyvesant High School in Manhattan where there was a

---

[1]Cohen uses S' instead of $\mathcal{D}(S)$. Also his definition of constructible set is slightly different from the one given here, but is equivalent to it. The equivalence follows from the fact that each L_α is *transitive*, meaning that each element of L_α is also one of its subsets.

[2]Gödel later used "V=L" instead of A for this statement, and Cohen does likewise. It should be mentioned that A also implies the axiom of choice and the so-called generalized continuum hypothesis $2^{\aleph_\alpha} = \aleph_{\alpha+1}$, so these are also consistent with the Zermelo-Fraenkel axioms.

[3]Reference [13], p. 153.

tradition of students challenging one another with difficult mathematics problems. He completed his Ph.D. at the University of Chicago, taking advantage of their willingness to to accept graduate students who had not completed a baccalaureate, by beginning his graduate studies there without first finishing his studies at Brooklyn College. He was a versatile and brilliant mathematician who made important contributions to a number of branches of mathematics. His great self-confidence in his abilities was sometimes seen as a kind of brashness. Attending the International Congress of Mathematicians in Moscow in 1966 (where he received the very prestigious Fields Medal for his work on CH), I heard him comment on the gross inefficiencies of the Communist system, remarking that any one of us taking charge of some aspect of that society could greatly improve things in a matter of weeks.

Undeterred by the general belief among professional logicians that Gödel's work had made the effort quixotic, Paul Cohen set out to prove the consistency of the Zermelo-Fraenkel axioms, in effect attempting to resurrect Hilbert's plan for the foundations of mathematics. It was only after he had given up on this endeavor that he turned his attention to the Continuum Hypothesis. Before Cohen's work, progress was stymied by the lack of tools for constructing models of the Zermelo-Fraenkel axioms. Cohen developed a new method he called *forcing* for obtaining such models that utterly revolutionized the subject of set theory. The method revealed a rich world of models that could be constructed to order by an adept in the technique. Cohen used forcing to show that neither Gödel's A nor CH could be proved from the Zermelo-Fraenkel axioms. Combined with what Gödel had shown using his constructible sets, this proved that both of these propositions are *independent* of the Zermelo-Fraenkel axioms. Others quickly learned the technique and proceeded to develop a dazzling array of new independence results.

The independence of CH leads to a difficult philosophical quandary. Is there still a fact of the matter to be resolved regarding CH, or has Cohen done all that can be done with the question? To *formalists* for whom set theory *is* the Zermelo-Fraenkel axioms and their consequences, the latter conclusion is evident. It is likely that most mathematicians, when they think about the matter at all, take this point of view. To those who still nourish the Brouwer-Weyl skepticism, Cohen's work simply reinforces their belief that CH itself is irretrievably vague. But there remain those who are convinced that there is a realm of sets in terms of which CH must be either true or false. To those, Cohen's work reinforces Gödel's emphasis on the weakness in the Zermelo-Fraenkel axioms. The models constructed by forcing are seen from the "outside" as thin wispy creatures by no means doing justice to Cantor's expansive vision. In such a model, sets that we can see have cardinality $\aleph_0$ are seen from "inside" the model as having different cardinalities simply because the one-one correspondences required to show that the sets have the same cardinality are not to be found inside the model. In the book Cohen suggests that CH may eventually be regarded as obviously false. Be that as it may, work on CH continues.

In writing this book, Cohen set himself the goal of developing his proof of the independence of CH without presupposing any knowledge of mathematical logic. He has accomplished this in less than 200 pages. En route he has covered the main methods and results of the subject. He did not aim for a polished work in which every last detail had been meticulously worked out. Rather he has presented the broad sweep of the subject underlining the intuitive ideas that more polished expositions sometimes hide. Dover Publications is doing a great service by making this master-work, long out of print, available again to the mathematical community.

Martin Davis
Berkeley, California
July 2008

My interaction with Kurt Gödel: The man and his work

PAUL J. COHEN
DEPARTMENT OF MATHEMATICS (EMERITUS)
STANFORD UNIVERSITY, UNITED STATES

On this centenary of Kurt Gödel's birth, it is most appropriate to celebrate the man who, more than any other, guided logic out of its philosophical past to become a vibrant part of present-day mathematics. His contributions are so well known, and the recognition he has received is so plentiful, that to recite them here would be extraneous. Rather, I have chosen to relate how first his work, and then my interaction with him, affected me so strongly.

Let me begin by explaining, and perhaps apologizing, for the somewhat personal tone of my remarks. Since the publication of Gödel's Collected Works, *with their very rich introductions to the various articles, as well as the historical material, I can add little of a purely biographical nature to Gödel's mathematical contributions. Thus, I shall have to speak mostly about what I know best: my own development and interest in Set Theory, how the example of Gödel's life work deeply affected me, the story of my own discoveries, and, finally, my personal interaction with Gödel when I presented my work to him.*

Introduction and Background

It was my great fortune and privilege to be the person who fulfilled the expectations of Gödel in showing that the Continuum Hypothesis (CH), as well as other questions in Set Theory, are independent of the usual set-theory axioms. Gödel foresaw this possibility in his article "What is Cantor's continuum problem?" (Gödel, 1947). Reading it greatly inspired me to work on the CH problem. Gödel's point of departure in discussing CH questions always seemed to be grounded in philosophical discussions and in the work of previous researchers. In my case, I felt it was necessary to start afresh and to treat the CH as an "ordinary" problem of mathematics, perhaps similarly to how Cantor, and later Hilbert, regarded it.

Here I would like to sketch briefly some of the evolution of my own ideas about the CH and their relation to Gödel's work, as well as his published opinions about my eventual resolution of the problem. None of my work, however, would have been possible without the momentous discovery of the *Constructible Universe,* which Gödel achieved around 1938. His approach in discovering the notion of a constructible set seems to have been decidedly influenced by the philosophical discussions of the past, especially the notion of impredicativity, which was emphasized by Poincaré and others. In this sense, I found the tone of his article difficult to understand, as it seemed to hover between philosophy and mathematics.

When I began my own attempts at solving the CH problem in 1962, I was strongly motivated by the idea of constructing a model. In this sense, my spiritual ancestor was perhaps Thoralf

Skolem more so than Gödel. Of course, I eventually realized that Gödel's "syntactical approach," where a constructible set could satisfy a certain property and constructible sets could be "collected" to form a model, was not very different. After my work was completed, I realized that a prejudice against models was widespread among logicians and even led to serious doubts about the correctness of my work. Nevertheless, as I was trying to construct interesting new models of Set Theory, I had to start with a given model of it. Thus, it seemed most natural to assume that this model was a collection of actual sets, where the membership relation has its usual meaning. Of course, the Löwenheim-Skolem Theorem guarantees the existence of such a model, even a countable one, but this cannot be proved within Zermelo–Fränkel Set Theory (ZF). Even so, its truth would intuitively be accepted by "everyone," even though technically it goes beyond ZF. On the other hand, Skolem specifically mentioned the problem of constructing more interesting models of Set Theory. In a remarkably prescient remark, he said that adjoining new sets of integers would be difficult, but probably of great interest. Although Skolem's remark was not known to me when I started my work, it did seem like a natural way to proceed. Thus, to violate the CH in this way, one would have to "adjoin" $\aleph$ 2 new sets of integers. This seemed an almost obvious way to proceed, so rather soon I found myself thinking in a manner similar to Skolem.

At first, I had not actually understood Gödel's work on the CH, finding the exposition of his monograph rather difficult. But as I thought about it more deeply, I realized that all my attempts were intimately allied to his methods, and I eventually mastered them completely. Suddenly, I had a tool with which I could make my own somewhat vague ideas more precise. Thanks to Gödel's discovery, I saw that the new universes of Set Theory I wished to construct consisted of sets that were constructible from a limited number of new sets, which I would introduce in a very particular way.

I would say that had the idea of constructing models by introducing new sets been established earlier, it would have been a natural development for Gödel to have achieved my own work. The fact that my methods are a kind of synthesis of the work of Gödel and Skolem has not been previously pointed out by authors; but I think it is fitting to honor the intuitions, fulfilled many years later, of these two pioneers.

A third theme of my work, namely the analysis of "truth," is made precise in the definition of "forcing." To people who read my work for the first time, this may very well strike them as the essential ingredient, rather than the two aspects of constructibility and the introduction of new sets, as mentioned above. Although I emphasized the "model" approach in my earlier remarks, this point of view is more akin to the "syntactical" approach, which I associate with Gödel. At least as I first presented it, the approach depends heavily on an analysis and ordering of statements and, by implication, on an analysis of proof.

Initially, I was not very comfortable with this approach; it seemed to be too close to philosophical discussions, which I felt would not ultimately be fruitful. In a sense, my enormous admiration for Gödel, and the fact that so much of his work was based on this kind of philosophical analysis, was unconsciously a driving factor in my persevering in this syntactical analysis. Of course, in the final form, it is very difficult to separate what is theoretic and what is syntactical. As I struggled to make these ideas precise, I vacillated between two approaches: the model theoretic, which I regarded as roughly more mathematical, and the syntactical-forcing, which I thought of as more philosophical. In this way, the body of Gödel's work, rather than particular results, helped me to believe that something could come from a syntactical analysis. I would say that my notion of forcing can also be viewed as arising from the analysis of how models are constructed, something in the spirit of the Löwenheim-Skolem Theorem.

If I were to comment on previous work that might have led to an earlier discovery of the independence results, these would be some of the salient points:

1. We have Skolem's emphasis on models. It seems that Gödel hardly discussed models of Set Theory. In particular, John Shepherdson's result about the impossibility of inner models to solve the problem should have received more attention (Shepherdson, 1951, 1952, 1953). Now, Gödel wrote no textbook on Set Theory, so we can only speculate as to what he would have had to say about models in general. His emphasis, as mentioned above, was syntactical, and thus he was concerned with consistency. The fact that a standard model for Set Theory (i.e., the Skolem Paradox) is a new axiom received scant attention. Of course, it is stronger than mere consistency.

2. It is easy to show that, assuming constructibility, no uncountable model that violates the CH can exist. Had this been realized, a key step in forcing—namely, the use of countable models—might have led to more rapid development of a solution.

3. In the spirit of Skolem, the addition of new sets was discussed briefly by András Hajnal and by Azriel Levy. I do not know whether this work intrigued Gödel to any extent.

My, perhaps, too dogmatic conclusion is that Gödel had no new method for a proof of independence, as he told some people. Yet, I cannot overestimate how the body of his work—his belief that a precise analysis of the axioms and consistency results was possible—played a crucial role in allowing me to go forward. As I shall now describe, these attitudes were reflected in my personal talks with Gödel after my work was completed.

Meeting Gödel

Soon after proving my first results, I had a great desire to meet Gödel and to do him and myself the great honor of personally explaining the proofs to him. I met him at Princeton with a preprint that gave all the essential details. He very graciously agreed to read it. After just a few days, he pronounced my proof correct. We had a series of intense discussions, and I asked whether he would communicate the paper to the *Proceedings of the National Academy of Sciences*, where his own results were first announced. He kindly consented to do this, and we had a lively correspondence about the best way to present it. He suggested several revisions. With this acknowledgment, and with his later overly kind remarks about my discovery, I felt that we shared a great bond in that we had successfully discovered, each in his own way, new fundamental methods in Set Theory.

I visited Princeton again for several months and had many meetings with Gödel. I brought up the question of whether, as rumor had it, he had proved the independence of the Axiom of Choice. He replied that he had, evidently by a method related to my own; but he gave me no precise idea or explanation of why his method evidently failed to succeed with the CH. His main interest seemed to lie in discussing the "truth" or "falsity" of these questions, not merely their undecidability. He struck me as having an almost unshakable belief in this "realist" position, which I found difficult to share. His ideas were grounded in a deep philosophical belief as to what the human mind could achieve. I greatly admired this faith in the power and beauty of Western culture, as he put it, and would have liked to understand more deeply the sources of his strongly held beliefs. Through our discussions, I came closer to his point of view, although I never shared completely his "realist" perspective that all questions of Set Theory were, in the final analysis, either true or false.

Let me turn now to a brief review of my intellectual development in mathematics.

High School—My Earliest Interest in Logic

In high school, number theory attracted me the most in mathematics, probably because of the simplicity of the statements of the results and the complexity and ingenuity of the proofs. I looked at the famous work of Edmund Landau, *Vorlesungen über Zahlentheorie (Lectures on Number Theory)* (Landau, 1927). The most complicated proof in the book was the famous result of Carl Siegel, generalizing the work of Axel Thue on approximation of algebraic numbers and applications to Diophantine equations. This proof is one of the rare, truly nonconstructive proofs in mathematics, and I was rather disappointed by this. My basic instinct was to prefer constructive proofs, of course, and so I began to think of exactly what I thought a constructive proof should be.

Thus, I was clumsily reconstructing elementary recursion theory, primitive recursive functions, etc. Simultaneously, results in partitions of integers led me to think how the results about sums of squares found by Jakob Jacobi and others could be generalized. As the early methods used infinite power series, I wondered whether—because we could answer all questions about polynomials by finite constructive methods—it would be possible to find (what I later learned are called) "decision procedures" for a larger class of problems.

At this time, I did not at all know Gödel's Incompleteness Theorem, which made this dream impossible. Eventually, I became dimly aware of Gödel's result, although not the precise statement. I felt it was probably primarily a theorem of philosophical interest, having little to do with the concrete questions of number theory that so infatuated me in those years.

Graduate School—The Continued Pull of Logic

When I started graduate school, for the first time I met students who had studied logic. I was told that the search for a decision procedure was probably doomed—although I did not clearly say exactly which class of problems concerning infinite power series most interested me. As fate would have it, Professor Stephen Kleene visited the University of Chicago, where I was enrolled, and he questioned me a bit about my "dream." In no uncertain terms, he told me that what I hoped to accomplish was impossible. So, I started to read his book on metamathematics (Kleene, 1952). There, in just a few pages, was a complete sketch of the Incompleteness Theorem. The rest of the book held little interest for me, and I was confused about many basic ideas. I still had a feeling of skepticism about Gödel's work, but skepticism mixed with awe and admiration.

I can say my feeling was roughly this: How can someone thinking about logic in almost philosophical terms discover a result that had implications for Diophantine equations? How very different it was from Landau's book. So, for a brief period, I believed that the famous Liar's Paradox, which is almost at the very heart of Gödel's reasoning, itself entailed a paradox—and even that his theorem, as I wanted to apply it, was false! But after a few days, I closed the book and tried to rediscover the proof, which I still feel is the best way to understand things. I totally capitulated. The Incompleteness Theorem was true, and Gödel was far superior to me in understanding the nature of mathematics.

Although the proof was basically simple, when stripped to its essentials I felt that its discoverer was above me and other mere mortals in his ability to understand what mathematics—and even human thought, for that matter—really was. From that moment on, my regard for Gödel was so high that I almost felt it would be beyond my wildest dreams to meet him and discover for myself how he thought about mathematics and the fount from which his deep intuition flowed. I could imagine myself as a clever mathematician solving difficult problems, but how could I emulate a result of the magnitude of the Incompleteness Theorem? There it stood, in splendid isolation and majesty, not allowing any kind of completion or addition because it answered the basic questions with such finality. I returned to the study of pure mathematics, having tasted a bit of the depth of logic. However, the rest of Kleene's book seemed rather philosophical, and, by temperament, even to the present day, I could not really be involved in philosophical controversy. So much seemed to depend on personalities and how one was able to express oneself with wit and elegance. As fate would have it, I was wrong in several regards.

Part of the standard curriculum at the university was a not-very-technical introduction to Set Theory, mostly the Axiom of Choice, facts about cardinality, etc. I saw these results as interesting, but merely as pieces of clever reasoning, inferior in that regard to the great body of mathematics that was attracting me—analysis, algebra, etc. As I have said, a career in logic seemed not very appealing because, in some sense, Gödel had pre-empted the entire field.

Thus, logic seemed to me to have little of the element of combinatorial thinking—or, more crudely put, cleverness—that I found in analysis and number theory. There was one exception, however. A small group of students were very interested in Emil Post's problem about maximal degree of unsolvability. I did dally with the thought of working on it, but in the end did not. Suddenly, one day a letter arrived containing a sketch of the solution by Richard Friedberg (Friedberg, 1957), and it was brought to my office. Amidst a certain degree of skepticism, I checked the proof and could find nothing wrong. It was exactly the kind of thing I would like to have done. I mentally resolved that I would not let an opportunity like that pass again. But, there seemed no way for me to find an entry into the kind of deep philosophical thinking that I so admired in Gödel and translate that into concrete mathematics.

The Beginnings of My Research Career

I wrote a thesis on harmonic analysis, on topics similar to those that led Georg Cantor to his discovery of Set Theory. Cantor's set-theory impetus was well known to me, but I saw no real connection with logic, at least not enough for me to begin to think about Set Theory as a possible research field. Thus, I essentially floated around the periphery of logic. Kleene's book contained essentially no mention of Gödel's result on the consistency of the CH, and I still had never actually seen a precise statement of the axioms of Zermelo and Fränkel. I was, however, aware of Skolem's Paradox concerning the existence of countable models of Set Theory. I was able to sketch a proof for myself, which seemed deep but not difficult. I might also mention that one of my office mates was Michael Morley, who wrote a very good thesis in Model Theory; but I knew nothing about it.

I then spent two years in Princeton at the Institute for Advanced Study and saw Gödel occasionally in the Common Room, but I had no contact with him. Princeton was an awe-inspiring place because of the famous people there, and Gödel was certainly one of them.

The Quest for Consistency

I then went to Stanford. At a departmental lunch, the conversation turned to the consistency of mathematics, mostly revolving around the ideas of Solomon Feferman (who is represented in this volume). At that point, something gelled, and I thought I saw a way to "prove" the consistency of mathematics by some kind of "construction." I gave a few lectures to a small audience, but I became discouraged in thinking that I was engaged in a futile enterprise—Gödel had shown that a consistency proof along the lines I had envisaged was impossible.

At some point, my attention turned to Set Theory and the CH. In Gödel's Princeton monograph, I read the precise statement of the axioms for the first time (Gödel, 1940). Essentially, the only weapon in my armory was the Löwenheim-Skolem Theorem, but I was convinced that the key question was constructing models of Set Theory. The Skolem construction offered little help. I did not see at first how it could be adapted to controlling the truth or falsity of given statements in the new model. It seemed that I had no real tools to work with—Set Theory was "too big" to handle, and Gödel's syntactical approach seemed more appealing for a while.

I read Gödel's article on Cantor's continuum problem (Gödel, 1947) before I had even considered attacking the problem of the CH. It was just of general interest for me, and to a large extent it was not even comprehensible to me. For one thing, it assumed a certain philosophical point of view and the philosophical ramifications of various attitudes toward it. I had never been attracted to philosophy, as I have explained, and thus I was not really aware of the various attitudes the mathematicians of earlier years had held.

Having said this, it is necessary to add that, of course, I had intuitive ideas of what constituted a correct proof in mathematics. In particular, in number theory I had begun to think about certain applications in which nonconstructive methods were used and was quite aware that these methods raised certain "problems." Of course, these were very far from the fundamental questions of Set Theory.

Did Gödel have unpublished methods for the CH? This is a tantalizing question. Let me state some incontrovertible facts. First, much effort was spent analyzing Gödel's notes and papers, and no idea has emerged about what kinds of methods he might have used. Second, I did ask him point blank whether he had proved the independence of CH, and he said no; but he had had success with the Axiom of Choice. I asked him what his methods were, and he said only that they resembled my own; he seemed extremely reluctant to give any further information.

My conclusion is that Gödel did not complete any serious work on this topic that he thought was correct. In our discussions, the word "model" almost never occurred. Therefore, I assume he was looking for a syntactical analysis that was in the spirit of his definition of constructibility. His total lack of interest in a model-theoretic approach quite astounded me. Thus, when I mentioned to him my discovery of the minimal model also found by John Shepherdson, he indicated that this was clear and, indirectly, that he knew of it. But he did not mention the implication that no purely "inner model" could be found. Since I also believe that he was strongly wedded to the syntactical approach, this would have been of great interest. My conclusion, perhaps uncharitable, is that he totally ignored questions of models and was perhaps only subconsciously aware of the minimal model.

The history of Gödel's involvement with the independence results, to be sure, has a somewhat sad conclusion. I think, without concrete evidence, of course, that it became interwoven in his mind with larger questions of the reality of Set Theory. His unshakable belief that all questions of Set Theory are "decidable" in some sense led him to think about the independence question in a somewhat confused way and eventually to a belief that he had discovered a method that would resolve the issue. Otherwise, I find no plausible reason why he would so adamantly refuse to discuss it with me. Of course, his general mental health was deteriorating in the last years, which no doubt was a contributing factor. It is not clear from biographical remarks of people who knew him at the time whether his breakdown in the early 1940s was a result of his frustration at not being able to solve the independence of CH.

Clearly Gödel's personal life was almost tragic in this period; and as far as the history of mathematics is concerned, I think no new information will arise.

Bibliography

Editor's note: The following bibliography was added after Paul Cohen's death to reflect sources of information that may be helpful to readers. It does not necessarily represent the sources that the author might actually have referred to while preparing this essay.

The following two references contain additional information on the discovery and reception of forcing, as well as further information on the author's intellectual development obtained from personal interviews by the writers; the second reference incorporates some material from the first:

Moore, C. H. (1988). The origins of forcing. In *Logic Colloquium '86*, ed. F. R. Drake and J. K. Truss. Amsterdam: North-Holland (Elsevier), pp. 143–73.

Yandell, B. H. (2001). *The Honors Class: Hubert's Problems and Their Solvers*. Wellesley, MA: A K Peters, 2001, pp. 59–84..

The following listing contains details about works relevant to this essay:

Church, A. (1968). Paul J. Cohen and the continuum problem. In *Proceedings of the International Congress of Mathematicians*. Moscow: International Congress of Mathematicians, pp. 15–20.

Cohen, P. J. (1963). A minimal mode for set theory. *Bulletin of the American Mathematical Society*, **69**, 537–40.

________ (1963). The independence of the Continuum Hypothesis. *I, Proceedings of the National Academy of Sciences*, **50**, 1143–48.

________ (1964). The independence of the Continuum Hypothesis. *II, Proceedings of the National Academy of Sciences*, **51**, 105–10.

________ (1965). Independence results in set theory. In *The Theory of Models: Proceedings of the 1963 International Symposium at Berkeley*. Amsterdam: North-Holland (Elsevier), pp. 39–54.

________ (1966). *Set Theory and the Continuum Hypothesis*, 4th edn. Reading, MA: Addison-Wesley.

Davis, M. (1965). *The Undecidable: Basic Papers on Undecidable Propositions, Unsolvable Problems, and Computable Functions*. Hewlett, NY: Raven Press.

Dawson, J. W., Jr. (1997). New light on the continuum problem. In *Logical Dilemmas: The Life and Work of Kurt Gödel*. Wellesley, MA: A K Peters, pp. 215–28.

Fränkel, A. A. (1919). *Einleitung in die Mengenlehre*. Berlin: Springer.

________ (1922). Zu den Grundlagen der Cantor-Zermeloschen Mengenlehre. *Mathematische Annalen*, **86**, 230–37.

________ (1922a). Der Begriff "definit" und die Unabhängigkeit des Auswahl-axioms. *Sitzungsberichte der Preussischen Akademie der Wissenschaften, Physikalisch-mathematische Klasse (The Concept "Definite" and the Independence of the Auswahl Axioms)*, pp. 253- 57. English translation in J. van Heijenoort (1967). *From Frege to Gödel: A Source Book in Mathematical Logic, 1879–1931*. Cambridge, MA: Harvard University Press, pp. 284–89.

_______ (1922b). Zu den Grundlagen der Mengenlehre. *Jahresbericht der Deutschen Mathematiker-Vereinigung (Angelegenheiten)*, **31**, 101–02.

_______ (1927). *Zehn Vorlesungen über die Grundlegung der Mengenlehre, gehalten in Kiel auf Einladung der Kant-Gesellschaft, Ortsgruppe Kiel, vom 8. –12. Juni 1925*. Leipzig: B. G. Teubner.

Fränkel, A. A., Bar-Hillel, Y., and Levy, A. (1973). *Foundations of Set Theory*, 2nd rev. edn. Amsterdam: North-Holland (Elsevier).

Friedberg, R. M. (1957). Two recursively enumerable sets of incomparable degrees of unsolvability (solution of Post's problem). *Proceedings of the National. Academy of Sciences*, **43**, 236–38.

Gödel, K. (1938). The consistency of the Axiom of Choice and of the Generalized Continuum Hypothesis. *Proceedings of the National. Academy of Sciences, USA*, **24**, 556–57.

_______ (1939). Consistency-proof for the Generalized Continuum Hypothesis. *Proceedings of the National. Academy of Sciences, USA*, **25**, 220–24.

_______ (1940). *The Consistency of the Axiom of Choice and of the Generalized Continuum Hypothesis with the Axioms of Set Theory*. Princeton: Princeton University Press.

_______ (1946). Remarks before the Princeton Bicentennial Conference on problems in mathematics. First published in M. Davis (1965). *The Undecidable: Basic Papers on Undecidable Propositions, Unsolvable Problems, and Computable Functions*. Hewlett, NY: Raven Press.

_______ (1947). What is Cantor's continuum problem? *American Mathematical Monthly*, **54**, 515–25.

_______ *Collected Works*, 5 vols., ed. S. Feferman, J. W. Dawson, Jr., S. C. Kleene, et al. New York: Oxford University Press:

Vol. I (2001): Publications 1929–1936.
Vol. II (2001): Publications 1938–1974.*
Vol. III (1995): Unpublished Essays and Lectures.
Vol. IV (2003): Selected Correspondence, A–G.
Vol. V (2003): Selected Correspondence, H–Z.

* Note that Volume II of the *Collected Works* contains reprints of all Gödel's published papers and monographs with extensive editorial comments. The original dates of publication have been listed above for historical accuracy.

Hajnal, A. (1956). On a consistency theorem connected with the Generalized Continuum Problem. *Zeitschrift für mathematische Logik und Grundlagen der Mathematik*, **2**, 131–36.

_______ (1961). On a consistency theorem connected with the Generalized Continuum Problem. *Acta Mathematica Academiae Scientiarum Hungaricae* (now *Acta Mathematica Hungarica)*, **12**, 321–76.

Hallett, M. (1984). Cantorian set theory and limitation of size. New York: Oxford University Press.

Hilbert, D. (1926). Über das Unendliche. *Mathematische Annalen*, **95**, 161–90. English translation in J. van Heijenoort (1967). *From Frege to Gödel: A Source Book in Mathematical Logic, 1879–1931*. Cambridge, MA: Harvard University Press, pp. 367–92.

_______ (1928). Die Grundlagen der Mathematik. *Abhandlugen aus dem mathematischen Seminar der Hamburgischen Universität*, **6**, 65–85. English translation in J. van Heijenoort (1967). *From Frege to Gödel: A Source Book in Mathematical Logic, 1879–1931*. Cambridge, MA: Harvard University Press, pp. 464–79.

Kleene, S. C. (1952). *Introduction to Metamathematics*. Amsterdam: North-Holland (Elsevier).

Landau, E. (1927). *Vorlesungen über Zahlentheorie (Lectures on Number Theory)*. Leipzig: Hirzel.

Levy, A. (1960). A generalization of Gödel's notion of constructibility. *Journal of Symbolic Logic*, **25**, 147–55.

_______ (1963). Independence results in set theory by Cohen's method. I, III, IV (abstract). *Notices of the American Mathematical Society*, **10**, 592–93.

_______ (1965). Definability in axiomatic set theory. *I. Logic, Methodology, and Philosophy of Science. Proceedings of the 1964 International Congress*, ed. Y. Bar-Hillel. Amsterdam: North-Holland (Elsevier), pp. 127–51.

Moore, G. H. (1982). *Zermelo's Axiom of Choice: Its Origins, Development, and Influence*. New York: Springer.

Mostowski, A. (1939). Über die Unabhängigkeit des Wohlordnungssatzes vom Ordnungsprinzip. *Fundamenta Mathematicae*, **32**, 201–52.

_______ (1950). Some impredicative definitions in the axiomatic set theory. *Fundamenta Mathematicae*, **37**, 111–24.

_______ (1952). On models of axiomatic systems. *Fundamenta Mathematicae*, **39**, 133–58.

Shepherdson, J. C. (1951). Inner models of set theory. *Journal of Symbolic Logic*, **16**, 161–90.

_______ (1952). Inner models of set theory. Part II. *Journal of Symbolic Logic*, **17**, 225–37.

_______ (1953). Inner models of set theory. Part III. *Journal of Symbolic Logic*, **18**, 146–67.

Skolem, T. (1923). Einige Bemerkungen zur axiomatischen Begründung der Mengenlehre. *Matematikerkongressen i Helsinfors den 4–7 Juli 1922, Den femte skandinaviska Matematikerkongressen, Redogörelse.* Helsinki: Akademiska Bokhandeln, pp. 217–32. English translation in J. van Heijenoort (1967). *From Frege to Gödel: A Source Book in Mathematical Logic, 1879–1931.* Cambridge, MA: Harvard University Press, pp. 290–301.

_______ (1930). Einige Bemerkungen zu der Abhandlung von E. Zermelo: Über die Definitheit in der Axiomatik. *Fundamenta Mathematicae*, **25**, 37–41.

_______ (1934). Über die Nicht-charakterisierbarkeit der Zahlenreihe mittels endlich oder abzählbar unendlich vieler Aussagen mit ausschließlich Zahlenvariablen. *Fundamenta Mathematicae*, **23**, 150–61.

Specker, E. (1957). Zur Axiomatik der Mengenlehre (Fundierungs- und Auswahl-axiom). *Zeitschrift für mathematische Logik und Grundlagen der Mathematik*, **3**, 173–210.

van Heijenoort, J. (1967). *From Frege to Gödel: A Source Book in Mathematical Logic, 1879–1931.* Cambridge, MA: Harvard University Press.

von Neumann, J. (1923). Zur Einführung der transfiniten Zahlen. *Acta litterarum ac scientiarum Regiae Universitatis Hungaricae Francisco-Josephinae, Sectio scientiarum mathematicarum*, **1**, 199–208. English translation in *From Frege to Gödel: A Source Book in Mathematical Logic, 1879–1931.* Cambridge, MA: Harvard University Press, pp. 346–54.

_______ (1925). Eine Axiomatisierung der Mengenlehre. *Journal für die reine und angewandte Mathematik*, **154**, 219–40. English translation in *From Frege to Gödel: A Source Book in Mathematical Logic, 1879–1931.* Cambridge, MA: Harvard University Press, pp. 393–413.

_______ (1929). Über eine Widerspruchsfreiheitsfrage in der axiomatischen Mengenlehre. *Journal für die reine und angewandte Mathematik*, **160**, 227–41.

Zermelo, E. (1908). Untersuchungen über die Grundlagen der Mengenlehre. *I., Mathematische Annalen*, **65**, 261–81.

_______ (1929). Über den Begriff der Definitheit in der Axiomatik. *Fundamenta Mathematicae*, **14**, 339–44.

________ (1930).Über Grenzzahlen und Mengenbereiche: Neue Untersuchungen über die Grundlagen der Mengenlehre. *Fundamenta Mathematicae*, **16**, 29–47.

________ (1932).Über Stufen der Quantifikation und die Logik des Unendlichen. *Jahresbericht der Deutschen Mathematiker-Vereinigung (Angelegenheiten)*, **41**, 85–88.

PREFACE

The notes that follow are based on a course given at Harvard University, Spring 1965. The main objective was to give the proof of the independence of the continuum hypothesis. To keep the course as self-contained as possible we included background material in logic and axiomatic set theory as well as an account of Gödel's proof of the consistency of the continuum hypothesis. Our review of logic is of necessity rather sketchy although we have tried to cover some of the fundamental concepts such as formal systems, undecidable statements and recursive functions. Actually, with the exception of the Löwenheim-Skolem theorem, none of the results of the first chapter are used in the later work and the reader who has had an introductory course in logic may omit this chapter. Its primary purpose is to accustom mathematicians who are not specialists in logic to the strictly precise point of view which is necessary when dealing with questions in the foundations of mathematics. Also, it is intended to clarify certain common confusions such as that of the concept of an undecidable statement in a particular axiom system with the concept of an unsolvable problem, which concerns methods of computation.

Since our very sincere hope is to make these notes intelligible to the large body of non-specialists who are interested in the problem, we have not adopted the very formalistic style which is to be found in some textbooks on logic. Rather we have tried to emphasize the intuitive motivations while at the same time giving as complete proofs as possible. No specific background is assumed, although we occasionally refer to examples from other parts of mathematics. Of course, it would be helpful if the reader were familiar with the development of "naive" set theory as it is customarily taught in courses on real variables or point set topology.

We would like to thank most heartily L. Corwin, D. Pincus, T. Scanlon, R. Walton, and J. Xenakis for taking notes for various sections and Jon Barwise for helping in the preparation of the final version. To Azriel Lévy we are deeply indebted for correcting many errors, for many improvements in presentation, and for helping to bring the entire stylistic level of these notes above their rather primitive original state. Thanks are also due to the members of the course at large who consistently challenged all mistakes and who by their stimulating discussion made teaching the course a distinct pleasure. The notes are certainly not in the polished form the subject warrants, but since there is no reasonably complete account of these questions in the literature, we thought it best to publish them in the present informal manner rather than to procrastinate indefinitely. Lastly, our sincere thanks go to the typists who worked on the notes, principally Sue Golan and Mari Wilson, for their fine efforts.

CONTENTS

CHAPTER I

GENERAL BACKGROUND IN LOGIC

1. INTRODUCTION

It is now known that the truth or falsity of the continuum hypothesis and other related conjectures cannot be determined by set theory as we know it today. This state of affairs regarding a classical and presumably well-posed problem must certainly appear rather unsatisfactory to the average mathematician. One is tempted to look more closely and perhaps more critically at the foundations of mathematics. Although our present "Cantorian" mathematics is highly successful in its treatment of abstractions, one must not overlook the fact that from the very beginning the use of infinite processes was regarded with suspicion by many people. In the 19th century, the objections regarding the use of convergent series and real numbers were met by Cauchy, Dedekind, Cantor and others, only to be followed by more profound criticisms from later mathematicians such as Brouwer, Poincaré and Weyl. The controversy which followed resulted in the formation of various schools of thought concerning the foundations. It is safe to say that no attitude has been completely successful in answering the fundamental questions, but rather that the difficulties seem to be inherent in the very nature of mathematics. Despite the fact that the continuum hypothesis is a very dramatic example of what might be called an absolutely undecidable statement (in our present scheme of things), Gödel's incompleteness theorem still represents the greatest obstacle to a satisfactory philosophy of mathematics. These fundamental difficulties, often dismissed by mathematicians, make the independence of the continuum hypothesis less surprising.

Gauss seems to have been the first mathematician to have expressed doubts about too free a use of infinities. In 1831, he wrote, "I protest ...against the use of an infinite magnitude as something completed, which is never permissible." Later, Kronecker expressed views which were critical of definitions that required an infinite process to verify that an object satisfied them. Cantor's work on set theory was the subject of much criticism to the effect that it dealt with fictions. Nevertheless, infinite sets are accepted today with few reservations. The traditional attitude accepts the construction of the real number system from the rationals as the last and final step in the long series of criticisms and re-examinations which have marked the history of mathematics. What possible objections can be raised to the construction of the real numbers? Simply this: although the reals are based on the integers, the vague notion of an arbitrary set of integers (or, equivalently, an arbitrary sequence of integers) must be introduced. Mathematicians inclined to a finitist point of view might hold that only sets which have an explicit rule to determine which integers are in the set should be admitted. For example, the school of Brouwer (Intuitionism) would only admit finite sets as legitimate objects of study, and even a single integer would not be considered defined unless a very definite rule for computing it was given. (For instance, the set consisting of 5 if Fermat's Last Theorem is true and 7 if it is false is not well-defined, according to Brouwer.) The criticism of Weyl and Poincaré was directed against "impredicative" definitions. Although their objections were not as extreme as Brouwer's, the acceptance of these criticisms would mean the destruction of large portions of mathematics.

Another source of objections was the paradoxes or antinomies of set theory. In Cantor's set theory a set was thought of as being defined by a property. Cantor himself pointed out that the set of all sets leads to an absurdity. Although this type of paradox (along with those of Russell, Burali-Forti and others), seems entirely remote from ordinary mathematical reasoning, the paradoxes did point out the necessity of extreme care when attempting to describe which properties describe sets. In 1908, Zermelo presented a formal set of axioms for set theory which encompassed all the present day reasonings in mathematics and yet which is presumably free from paradoxes. This axiomatization of set theory was in keeping with the spirit of the school of Formalism, led by David Hilbert. According to the Formalist point of view, mathematics should be regarded as a purely formal game played with marks on paper, and the only requirement this game need fulfill is that it does not lead to an inconsistency. To completely describe the game required setting down the rules of mathematical logic with much greater precision than had been previously done. This was done, and the Formalists turned their attention to showing that various systems were consistent. As is well known, this hope was destroyed by Gödel's discovery of the incompleteness theorem, which implies that the consistency of a mathematical system cannot be proved except by methods more powerful than those of the system itself.

Despite this failure, the Formalist program contributed greatly to the development of logic by establishing a systematic study of mathematical systems. In these notes, our first object will be to describe how a mathematical system can be completely reduced to a purely formal game involving the manipulation of symbols on paper. By a <u>formal system</u> we shall mean a finite collection of symbols and perfectly precise rules for manipulating these symbols to form certain combinations called "theorems". Of course, these rules must be given in informal mathematical language. However, we shall demand that they be completely explicit rules requiring no infinite processes to check and that in principle they can be coded into a computing machine. In this way questions concerning infinite sets are replaced by questions concerning the combinatorial possibilities of a certain formal game. Then we will be able to say that certain statements are not decidable within given formal systems.

2. FORMAL LANGUAGES

If we examine Peano's axioms for the integers, we find that they are not capable of being transcribed in a form acceptable to a computing machine. This is because the crucial axiom of induction speaks about "sets" of integers but the axioms do not give rules for forming sets nor other basic properties of sets. Here is an example of the difficulty in satisfying the stringent requirements for a formal system outlined above. When we do construct a formal system corresponding to Peano's axioms we shall find that the result can not quite live up to all our expectations. This difficulty is associated with any attempt at formalization.

Two types of symbols will appear in our formal language. First, there are those symbols common to all mathematical systems. Then there are those used to denote particular concepts in special branches of mathematics, such as the symbols for addition, group multiplication, adjoint of a matrix, etc. The general symbols we shall use are the following:

$\sim$	&	$\vee$	$\rightarrow$	$\leftrightarrow$
not	and	or	implies	if and only if

$\forall$	$\exists$	$=$	(,)	x, '
for all	there exists	equals	parentheses	variable symbols

The words beneath the symbols have, in principle, nothing to do with our formal language. Rather, we are thinking of the symbol "$\leftrightarrow$" as formally representing the words "if and only if". The ordinary meanings of these words will suggest certain rules concerning the symbols, but the formal game must be played by the rules without recourse to any meanings which may have suggested them.

The first five symbols are known as propositional connectives. The symbols which formally represent "for all" and "there exist" are known as universal and existential quantifiers respectively. Parentheses are used in the formation of expressions to insure their unique readability. In any given discussion one may need arbitrarily many variables so that we shall use $x, x', x'', \ldots,$ as symbols for variables. In practice we will use the letters $x_1, x_2, x, \ldots$ or x, y, z although in theory these should be replaced by $x, x', x'', \ldots$. In this way we need only a finite set of symbols. Our list of symbols is in no sense the most economical for our purposes since (as will follow from the rules for our formal system) some of the symbols can be avoided by combinations of others. For example, it will follow from rules A and G below that $\forall$ could be replaced by $\sim \exists \sim$. We shall sometimes omit parentheses or employ other abbreviations in our formulas if there is no danger of confusion.

To express interesting statements we need special symbols which will be used to formally represent particular relations under consideration. For this purpose we admit a finite set of relation symbols R_1, $R_2, \ldots$. To each R_i is assigned an integer $n_i \geq 1$, n_i intuitively indicating that the relation represented by R_i is a relation between n_i objects. For example, if $n_1 = 2$, then R_1 is a binary relation (such as $\leq$ between real numbers), and in our rules for forming formal expressions we will require that R_1 be followed by two variables or constants as in $R_1(x,y)$ and $R_1(c,z)$. We will say that R_i is an n_i-ary predicate symbol, and sometimes write $R_i(x_1,\ldots,x_{n_i})$ to indicate n_i explicitly. We shall also use particular symbols to represent constants which play a special role in the system. For example, in group theory it is convenient to have a symbol for the identity element. We use the letter "c" together with " ' " to generate the symbols c, c', c", etc. to represent constants. In practice, we write more simply c_1, c_2, etc.

Before we give the precise rules for forming the formal expressions of our formal language we give some examples. Instead of thinking of "addition" an an operation on pairs of numbers, we can think of it as a relation between triples of numbers, namely the relation

$$r + s = t.$$

Thus if we are considering a formal language in which R_1 is a ternary relation symbol, we could let R_1 represent this relation. Then the uniqueness of addition would be insured by

$$\forall x\ \forall y\ \forall z\ \forall u\ ((R_1(x,y,z)\ \&\ R_1(x,y,u)) \rightarrow z = u.$$

If we let multiplication $(r \cdot s = t)$ be represented by $R_2(x,y,z)$, then the associativity of multiplication becomes

$$\forall x\ \forall y\ \forall z\ \exists t_1\ \exists t_2\ \exists t_3 (R_2(x,y,t_1)\ \&\ R_2(t_1,z,t_2)\ \&\ R_2(y,z,t_3)\ \&\ R_2(x,t_3,t_2))$$

The existence of an additive identity can be stated as $\exists x\ \forall y (R_1(x,y,y))$ but this would be inconvenient in many cases. For, whenever an argument used this additive identity we would have to restate its existence. It is simpler to use a constant symbol "0" and introduce as an axiom

$\forall x\ (x + 0 = x)$. In general the use of abbreviations is indispensable for making mathematics comprehensible and we shall not hesitate to use them when their meaning is clear and when it is also clear how to replace them by the strict formal language.

Let us examine further our formal system using only the special relation symbols R_1 and R_2 representing addition and multiplication. It is not hard to see that we can express all the axioms of Field Theory within the system. (Observe that we are not yet discussing proofs or "true" statements, merely what can be expressed.) More complicated assertions such as the fact that a quadratic equation has at most two roots can be written as follows:

$$\forall a,b\ \exists u\ \exists v\ \ \forall x\ ((x^2 + ax + b = 0) \rightarrow (x = u \vee x = v)).$$

We leave as an exercise to show how the formula $x^2 + 2x + b = 0$ can be transcribed in the formal system. One encounters difficulty if one tries to say that an equation of degree n has at most n roots. This is because we have no notation (within the formal system) for an integer to denote the degree of an arbitrary polynomial nor an induction procedure to define the notion of a polynomial of arbitrary degree. Thus we cannot handle properties which involve the notion of an arbitrary integer. Nevertheless, one could formulate the theorems of Galois theory for quadratic, cubic, etc. extensions by speaking only of root permutations. Galois theory in its customary formulation speaks about sets, such as sub-fields, sub-groups, etc. and so cannot be expressed directly in our system.

We now give the precise rules for forming grammatically correct statements. These are called well-formed formulas (wff). Recall that our formal language consists of the general symbols given above including variable symbols and constant symbols, as well as a finite number of relation symbols each associated with an integer.

Rules for well-formed formulas

1. $x = y$, $x = c$, $c = c'$ are wff's where x and y are variable symbols and c and c' are any constant symbols.
2. If R is an n-ary relation symbol and each of $t_1,\ldots,t_n$ is either a variable or constant symbol, then $R(t_1,\ldots,t_n)$ is a wff.

3. If A and B are wff's so are $\sim(A)$, (A) & (B), $(A) \vee (B)$, $(A) \rightarrow (B)$, and $(A) \longleftrightarrow (B)$.

4. If A is a wff, then so are $\exists xA$ and $\forall xA$.

Observe that our wff's may have loose constant and variable symbols dangling about. Also Rule 4 allows an expression such as $\forall xy = z$ as a wff. This is to be interpreted intuitively as meaning that if $\forall x$ or $\exists x$ occurs before a wff A which does not involve x then the effect of the quantifiers is nil, and they can be omitted. This brings us to the precise distinction between bound and free variables.

Definition. Each occurrence of a variable symbol in a wff is defined as free or bounded as follows:

1. Every variable occurring in a formula of the form mentioned in Rules 1 and 2 is free.

2. The free and bound occurrences of variables in the wff's mentioned in Rule 3 are precisely the same as those for A and B separately.

3. The free and bound occurrences of a variable in a formula $\exists xA$ or $\forall xA$ are the same as those of A except that every free occurrence of x is now considered bound.

The reader may object that we have been too careless in Rule 4 by allowing bound variables to be bound again. Actually, no confusion is possible if we agree that such quantification has no effect, but to be more specific we can make the following definition:

Definition. A wff is called "good" if in the application of Rule 3, A and B have only free variables in common, and in the application of Rule 4, x occurs as a free variable in A.

The conventions concerning the use of parentheses are such that no ambiguity can arise. This would actually require a simple combinatorial argument, which we omit.

Definition. A statement is a formula with no free variables.

3. UNIVERSALLY VALID STATEMENTS

The object of mathematics is to discover "true" theorems. We shall use the term "valid" to describe statements formed according to certain rules and then shall discuss how this notion compares with the intuitive idea of "true". When we write down a statement involving certain constant and relation symbols, we have not indicated any specific interpretation of these symbols. If such a statement is to be intuitively considered as "true" it must be true independently of how these symbols are interpreted. For example, if A is the conjunction of the usual axioms for field theory and B is the statement that a quadratic equation has at most two roots, then $A \to B$ is a true statement, because in any system where the axioms hold with regard to two ternary relations (which we may call addition and multiplication), a quadratic equation may have at most two roots. We shall introduce the concept of a model for a formal system and a fundamental result will be the identification of the valid statements with those that are true in every model.

The rules for forming valid statements are known as the Predicate Calculus. A certain simple sub-system of these rules, whose importance was realized before the other rules, is known as the Propositional Calculus. These are the rules concerning the manipulation of the symbols $\sim$, &, $\vee$, $\to$, $\longleftrightarrow$. For the moment, let $A_1, A_2, \ldots$ be variable letters, not to be confused with the variables used in the formal language. These variables will eventually be replaced by wff's.

Definition. A propositional function is a formal string of symbols defined as follows:

1. If A is a variable letter then A is a propositional function.
2. If P and Q are propositional functions so are $\sim(P)$, $(P) \& (Q)$, $(P) \vee (Q)$, $(P) \to (Q)$, and $(P) \longleftrightarrow (Q)$.

If P is a propositional function of the variables $A_1, \ldots, A_n$ (that is, P only involves variable letters among the $A_1, \ldots, A_n$, and not necessarily all of these), we wish to indicate how the truth or falsity of P depends on that of A_i. To this end we associate with P a function defined on the set of all n-tuples of 0 and 1 (i.e., $(\epsilon_1, \ldots, \epsilon_n)$, where ϵ_i is either 0 or 1) taking the values 0 and 1. This is done as follows:

1. The propositional function A_i corresponds to the projection function $(\epsilon_1, \ldots, \epsilon_n) \rightarrow \epsilon_i$.

2. If f is the function corresponding to P and g that for Q, the functions corresponding to $\sim(P)$, $(P) \,\&\, (Q)$, $(P) \vee (Q)$, $(P) \rightarrow (Q)$, $(P) \longleftrightarrow (Q)$ are respectively, $\varphi_1(f)$, $\varphi_2(f,g)$, $\varphi_3(f,g)$, $\varphi_4(f,g)$, $\varphi_5(f,g)$ where the φ_i are given by the "truth tables":

x ↓	
0	1
1	0

$\varphi_1(x)$

$\sim$

x ↓ \ y →	0	1
0	0	0
1	0	1

$\varphi_2(x,y)$

&

x ↓ \ y →	0	1
0	0	1
1	1	1

$\varphi_3(x,y)$

$\vee$

x ↓ \ y →	0	1
0	1	1
1	0	1

$\varphi_4(x,y)$

$\rightarrow$

x ↓ \ y →	0	1
0	1	0
1	0	1

$\varphi_5(x,y)$

$\longleftrightarrow$

The reader will easily check that this corresponds to the usual usage of the propositional connectives, with the understanding that 1 represents truth and 0 falsity. We define a propositional function as identically true if the corresponding function takes only the value 1. This means intuitively that the propositional function is "true" no matter what statements are substituted for the variables. We can now state the basic rule of the Propositional Calculus:

<u>Rule A</u> (<u>Rule of the Propositional Calculus</u>)

If P is a propositional function of the variable letters $A_1, \ldots, A_n$ which is identically true, then the result of replacing each A_i by any statement is a <u>valid</u> statement.

Recall that our intended meaning of the term "valid statement" will be a statement which is intuitively true in any interpretation of the relation and constant symbols used to form it. There is another rule of the predicate calculus related to A which allows us to form new valid statements from old ones. It has often been illustrated by statements concerning the mortality of Socrates.

<u>Rule B</u> (<u>Rule of Inference</u>)

If A and $(A) \rightarrow (B)$ are valid statements so is B.

We next give the rules for manipulating the sign of equality.

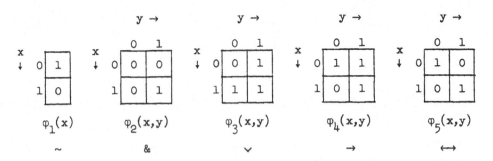

Rule C (Rules of Equality)

1. $c = c$, $(c = c') \to (c' = c)$, and $((c = c') \mathbin{\&} (c' = c'')) \to (c = c'')$ are valid statements where c, c', c'' are any three constant symbols.

2. If A is a statement, c and c' constant symbols, and if A' represents A with every occurrence of c replaced by c', then $(c = c') \to ((A) \to (A'))$ is a valid statement.

Rule D (Change of Variables)

If A is any statement and A' results from A by replacing each occurrence of the symbol x with the symbol x', where x and x' are any two variable symbols, then the statement $(A) \longleftrightarrow (A')$ is a valid statement.

Both of these rules are obvious and require no discussion. It is a simple exercise to show that for every statement A, there is a "good" statement A' obtained by replacing the variable symbols in A by other variable symbols, and such that $(A) \longleftrightarrow (A')$ is a valid statement. For the next rule, let $A(x)$ represent a formula with one free variable x and in which every occurrence of x is free and let $A(c)$ represent the result of replacing every occurrence of x by the constant symbol c.

Rule E (Rule of Specialization)

$(\forall x \ A(x)) \to (A(c))$ is a valid statement where c is any constant symbol.

The next rule is a bit misleading and requires some explanation. Often in arguments we say "let c be an arbitrary but fixed integer". We then proceed to reason about c and come to a certain conclusion $A(c)$. We can then deduce that $\forall x \ A(x)$ since we used no special properties of c. In reality, we have treated c as a variable, even though we called it a constant. This is because our valid statements will be true in every interpretation of the constant and relation symbols. We could express this by: if $A(c)$ is a valid statement so is $\forall x \ A(x)$. However, rule F is in a form more convenient for our purposes and yields the above, as we shall see.

Rule F

Let B be a statement not involving c or x. Then if $A(c) \to B$ is valid, so is $\exists x A(x) \to B$.

The next rule will allow us to bring every formula involving quantifiers into a form in which it begins with a quantifier.

Rule G

Let $A(x)$ have x as the only free variable and let every occurrence of x be free. Let B be a statement which does not contain x. Then the following are valid statements.

$$(\sim(\forall x A(x))) \leftrightarrow (\exists x \sim (A(x)))$$

$$((\forall x A(x)) \,\&\, (B)) \leftrightarrow (\forall x((A(x)) \,\&\, (B)))$$

$$((\exists x A(x)) \,\&\, (B)) \leftrightarrow (\exists x(A(x)) \,\&\, (B))$$

Definition. Let S be a collection of statements. We say that A is derivable from S, if for some $B_1,\ldots,B_n$ in S, the statement $((B_1) \,\&\, \ldots \,\&\, (B_n)) \rightarrow (A)$ is valid.

Fact: If S and S' are collections of statements such that every statement in S' is derivable from S, then every statement derivable from S' is derivable from S.

Proof. Exercise. In the course of a proof we often reason by contradiction or assume certain hypotheses temporarily. Such steps are all justified by the propositional calculus, although we shall not give all the formal details in applications. Assume for example we know that $A(c)$ is a valid statement. We shall show that $\forall(x)\, A(x)$ is valid. (This is not the same as saying $A(c) \rightarrow \forall(x)\, A(x)$ is valid, which it is not in general.) It is sufficient to show that $\sim \forall(x)\, A(x)$ leads to a contradiction (i.e., $B \,\&\, \sim B$ for any B). Equivalently by Rule G we show that $\exists x \sim A(x)$ leads to a contradiction. But $\sim A(c)$ does lead to a contradiction since $A(c)$ is valid and hence by Rule F so does $\exists x \sim A(x)$.

4. GÖDEL COMPLETENESS THEOREM

Having now given rules for forming valid statements we come to the problem of identifying these statements with the intuitively "true" statements. This discussion will be carried out in the spirit of traditional mathematics, that is to say, outside of any formal language. We shall use some elementary notions of set theory. After we have formalized set theory itself, then of course this discussion can be expressed in that formal system. In our original discussion, we had a finite number of symbols. This was important for foundational purposes, in order to reduce mathematics to a formal game playable by a computing machine. However, for some purposes it is of interest to allow arbitrarily many constant and relation symbols. We write the proofs of Sections 4 and 5 for this more general case.

Assume now that we are dealing with a collection S of statements involving constants c_α, $\alpha \in I$, and relation symbols R_β, $\beta \in J$, where each R_β has a fixed number of variables. Let M be a non-empty set and let $c_\alpha \to \bar{c}_\alpha$ be a map from the constant symbols to elements of M, not necessarily distinct, and $R_\beta \to \bar{R}_\beta$ a map which associates to a k-ary relation symbol, a subset of the k-fold direct product, $M \times M \times \cdots \times M$. We shall then say that we have an interpretation of the constants c_α and the relation symbols R_β in the set M. To every statement using only these constant symbols and relation symbols, we shall associate its "truth value" under this interpretation. Intuitively, of course, we merely mean whether or not the statement is true in M under the given interpretation of the constant and relation symbols. However, a precise definition is easy to give if we proceed by induction on the length of formulas.

Definition. Let A be a formula with free variables among $x_1,\ldots,x_n$, $n \geq 0$, and let $\bar{x}_1,\ldots,\bar{x}_n$ be elements of M. We define the truth value of A (in M) at $\bar{x}_1,\ldots,\bar{x}_n$.

1. If A is of the form $x_i = x_j$, $x_i = c$, or $c_i = c_j$, then A is true at $\bar{x}_1,\ldots,\bar{x}_n$ if $\bar{x}_i = \bar{x}_j$, $\bar{x}_i = \bar{c}$, or $\bar{c}_i = \bar{c}_j$, respectively.

2. If A is $R(t_1,\ldots,t_m)$ where R is an m-ary relation symbol and each t_i is a constant symbol or one of the $x_1,\ldots,x_n$, then A is true at $\bar{x}_1,\ldots,\bar{x}_n$ if the m-tuple $\langle \bar{t}_1,\ldots,\bar{t}_m \rangle$ is in $\bar{R}$ (the subset of M^m associated with R under the given interpretation).

3. If A is a propositional function of formulas, we evaluate the truth of A at $\bar{x}_1,\ldots,\bar{x}_n$ by means of the propostional calculus.

4. If A is of the form $\forall(y)B(y,x_1,\ldots,x_n)$[resp. $\exists yB(y,x_1,\ldots,x_n)$] then A is true at $\bar{x}_1,\ldots,\bar{x}_n$ if, for all $\bar{y}$ in M [resp. for some $\bar{y}$ in M] $B(y,x_1,\ldots,x_n)$ is true at $\bar{y}$, $\bar{x}_1,\ldots,\bar{x}_n$.

To avoid any ambiguity over substitution of variables we may assume that all the formulas are good wff's as defined in Section 3. We note that if A is a statement we can take $n = 0$, and our definition is just truth in M under the given interpretation.

<u>Definition</u>. If S is a set of statements containing c_α and R_β, M a set and $c_\alpha \to \bar{c}_\alpha$ and $R_\beta \to \bar{R}_\beta$ are maps as above, we say that M is a model for S (under the interpretation) if all the statements of S are true in M.

Strictly speaking a model is a set M together with an interpretation of some (possibly none) constant and relation symbols. Unless it is necessary for clarity, we shall suppress any mention of the explicit interpretation.

<u>Definition</u>. A set of statements is said to be consistent if the statement $A \,\&\, \sim A$ cannot be derived from S for any A.

The point of these definitions is the following obvious fact:

THEOREM 1. If A is a valid statement, it is true in every model. If a set of statements S has a model then it is consistent.

We omit the tedious proof, which is merely a verification that the rules of the predicate calculus correspond to correct methods of deduction. A much more interesting question is whether the rules we have given exhaust all possible deductions. With out notion of models we can phrase this quite precisely.

THEOREM 2. <u>Gödel Completeness Theorem</u>. Let S be any consistent set of statements. Then there exists a model for S whose cardinality does not exceed the cardinality of the number of statements in S if S is infinite, and is countable if S is finite.

The proof uses the axiom of choice (unless the set S is already well-ordered). Given S we shall explicitly show how to construct a model for S. The proof however is non-constructive in the sense that the construction of the model for S may depend upon examining an infinite number of possibilities. Nevertheless, if S is finite, the model can be taken as the set of integers and the resulting relations will be arithmetically definable (i.e, definable by formulas using only addition and multiplication). This will follow from the form of the proof although we shall not go into the matter.

The first step is an examination of the special case in which S does not contain any quantifiers.

THEOREM 3. (Completeness of the Propositional Calculus). If S contains no quantifiers and is consistent, then there is a model M for S in which every element of M is of the form $\bar{c}_\alpha$ for some c_α appearing in S.

We need one lemma.

LEMMA. If T is a consistent set of statements, A an arbitrary statement, either $T \cup \{A\}$ or $T \cup \{\sim A\}$ is consistent.

Proof. If $T \cup \{A\}$ is inconsistent, then for some B_i in T, $A \,\&\, B_1 \,\&\, \cdots \,\&\, B_n \to C \,\&\, \sim C$ for some C, is valid. If $T \cup \{\sim A\}$ is inconsistent, then for some B_i' in T, $(\sim A) \,\&\, B_1' \,\&\, \cdots \,\&\, B_m' \to C \,\&\, \sim C$ is valid. The propositional calculus now implies that $B_1 \,\&\, \cdots \,\&\, B_n \,\&\, B_1' \,\&\, \cdots \,\&\, B_m' \to C \,\&\, \sim C$ is valid, so that T must be inconsistent.

Now to prove Theorem 3. Let S be well-ordered. This induces a well-ordering on all the constant and relation symbols which appear in S. This in turn induces a well-ordering of all possible statements of the form $c_i = c_j$ and $R_\beta(c_1, \ldots, c_n)$ where c_i and R_β are constant and relation symbols occurring in S. Call these statements F_α. We now define statements G_α by induction on α. If F_α is consistent with $S \cup \{G_\beta | \beta < \alpha\}$ we put $G_\alpha = F_\alpha$, otherwise put $G_\alpha = \sim F_\alpha$. By our lemma and by induction on α, it follows that $S \cup \{G_\beta | \beta \leq \alpha\}$ is consistent for all α. Since any contradiction must be derived from a finite number of statements, we see that $H = S \cup \{G_\alpha\}$ is a consistent system. For each c_α occurring in S, define $\bar{c}_\alpha = c_\beta$ where β is the least index such that $c_\alpha = c_\beta$ belongs to H. (There must be some such since $c_\alpha = c_\alpha$ belongs to H.) Let M be the set of $\bar{c}_\alpha$, and define $\bar{R}_\beta$ as the set of all $\langle \bar{c}_{\alpha_1}, \ldots, \bar{c}_{\alpha_n} \rangle$ such that $R_\beta(c_{\alpha_1}, \ldots, c_{\alpha_n})$ is in H. Thus, our model consists of a subset of the formal symbols c_α. An examination of Rule C will show that if there is any ambiguity in our determination of the relations $\bar{R}_\beta$ on M, then there must be a contradiction in H. It is also easy to see that every statement in H, or its negation, must be a consequence of the G_α since every atomic relation occurring in a statement of S or its negation appears among the G_α. A negation of a statement in S cannot be such a consequence since in that case H would not be consistent. Since we have defined M so that all the G_α are true, it follows that in M all the statements of H and hence of S are also true. Thus Theorem 3 is proved.

So far, we have only used Rules A, B, and C. To prove Theorem 2 we shall need the other rules of the predicate calculus as well. The basic idea is to replace the original system S by a system without quantifiers, such that the model for the new system given to us by Theorem 2 will also be a model for S. To do this, we shall systematically introduce new constants as they are demanded by the system S. More precisely, let T be a system of statements each of which either has no quantifiers or begins with a quantifier. Using Rules D and G, we see that this is not an essential restriction. We form a new system as follows: If the statement $\forall(x)\, A(x)$ and the symbol c occur in T adjoin the statement $A(c)$. Also for each statement of the form $\exists x A(x)$ choose a new constant symbol c, not already used in T, and adjoin the statement $A(c)$. The resulting system we call T^*. Note that T is a subset of T^*.

LEMMA. If T is consistent so is T^*.

The proof is an immediate consequence of Rules E and F. For, the first type of statement adjoined to T to form T^* is necessarily a consequence of T. As for the second type, Rule F shows that if a contradiction can be reached with the adjoined statement it can be reached without it. This is the reason we took Rule F in the form we did.

As noted above, Rules D and G allow us to replace an arbitrary system T with an equivalent one for which we can define T^*, so that we can consider T^* as always defined. For the consistent system S, we now define S_n for each integer n by putting $S_0 = S$ and $S_{n+1} = S_n^*$.

Put $\bar{S} = \bigcup_n S_n$ and let M be the model given by Theorem 2 for the subset of $\bar{S}$ consisting of those statements without quantifiers. Let A be a statement in $\bar{S}$ inolving r quantifiers. Assume that any statement in $\bar{S}$ with fewer than r quantifiers is true in M. Clearly if r is zero, our assumption is fulfilled. If r is not zero, assume that A is of the form $\forall(x)\ B(x)$. Any element in M must be of the form $\bar{c}_\alpha$ where c_α is one of the constant symbols occurring in S. Since the S_n are increasing, both A and c_α are in some S_k. Then $B(c_\alpha)$ occurs in S_k^*, hence in $\bar{S}$, and has fewer than r quantifiers. So, for any $\bar{c}_\alpha$ in M, $B(c_\alpha)$ is true in M, and hence $\forall(x)\ B(x)$ is true in M. The case in which A is of the form $\exists xB(x)$ is handled similarly. It is also clear that the cardinality of T* for any system T, is the same as that of T if T is infinite and otherwise is finite. This clearly yields the conclusion in the theorem concerning the cardinality of M.

The Gödel completeness theorem now assures us that no essential rule of deduction has been omitted from the predicate calculus. We state this as a corollary.

COROLLARY. If A is not derivable from S, then there is a model for S in which A is false.

<u>Proof</u>. $S \cup \{\sim A\}$ is a consistent system and hence has a model.

COROLLARY. If a statement is true (i.e., true in every model), then it is provable.

<u>Proof</u>. Take S to be empty in the previous corollary.

The completeness of the predicate calculus is in sharp contrast with the incompleteness of the axioms for mathematics, as will be explained in a later section. However, the completeness theorem also shows some crucial limitations of the notion of a formal system as is indicated by the following corollary.

COROLLARY. If S admits an infinite model or even arbitrarily large finite models, then S admits models of arbitrarily large cardinalities.

<u>Proof</u>. Let c_α be an arbitrary set of new constant symbols and adjoin to S the equations $\sim c_\alpha = c_\beta$ for all $\alpha \neq \beta$ to form a new

system T. The system T is consistent. For, any contradiction must be a consequence of finitely many statements of the form $\sim c_\alpha = c_\beta$ together with the system S. But since S has models in which we can find arbitrarily large finite sets of distinct constants, and since T puts no further restrictions on these c_α, we see that this subsystem of T has a model and so is consistent, which means that T itself is consistent. Thus T has a model and hence S has a model whose cardinality is at least that of the set of c_α introduced and the corollary is proved. This corollary shows that no system of axioms can have a unique model (up to isomorphism) unless this unique model is finite. So one sees that the usual systems of mathematics such as the integers and real numbers, which presumably have only one model, cannot be described completely by <u>any</u> formal system of axioms. As mentioned, if one looks at the classical axioms for these systems, one sees that they are not formal systems in our sense, since they speak informally about the set concept. Later the incompleteness theorem will show us that in an even more modest sense of the word, the axioms for mathematics cannot be complete. We close with one more corollary.

COROLLARY (compactness theorem). If every finite subset of a system S has a model, then S has a model.

5. THE LÖWENHEIM-SKOLEM THEOREM

A very interesting feature of the completeness theorem is the information concerning the cardinality of the models whose existence it asserts. It implies for example that however we formalize set theory with all its implications concerning the existence of sets of large cardinality, the set of axioms we present will have a countable model. Since the construction of the model was rather indirect, we do not expect that the objects of the model be identifiable with "real" sets, nor that the one undefined relation, in this case the relation of membership, be identifiable with the "real" membership relation. That is, in the case of set theory, or number theory, or the real number system, we have in mind one particular model, and we are primarily interested in it, or perhaps in sub-models of it. We shall now be concerned with the relations which can exist between models of the same formal system.

Definition. Let M_1 and M_2 be two models for a given formal system, i.e., we are given for each, maps from certain constant symbols c_α into the model, and maps associating with certain relation symbols R_β, certain subsets of n-tuples of M_1 and M_2. We say that M_1 and M_2 are elementarily equivalent if the statements in the formal language which are true in M_1 are precisely those true in M_2. If $M_1 \subseteq M_2$ we shall say that M_1 is an elementary sub-model of M_2, if the constants $\bar{c}_\alpha$ are the same in M_2 as in M_1, the relations $\bar{R}_\beta$ on M_1 are the restrictions of the relations $\bar{R}_\beta$ on M_2, and for every formula $A(x_1,\ldots,x_n)$, if $\bar{x}_i$ are in M_1 then $A(x_1,\ldots,x_n)$ is true in M_1 at $\bar{x}_1,\ldots,\bar{x}_n$ if and only if it is true in M_2 at $\bar{x}_1,\ldots,\bar{x}_n$.

The notion of elementarily equivalent systems bears some relation to that of isomorphic systems, the latter of course being a much stronger concept. It has not found too much application in mathematics, except in the theory of real-closed fields, and more recently in p-adic number fields. The condition for an elementary sub-model is stronger than that for elementary equivalence because it allows statements which involve specific individuals in M_1 other than the constants. We are now ready to state a theorem whose proof is almost identical with that of the completeness theorem and which states a similar conclusion. It was found earlier than the completeness theorem and was perhaps the first genuine theorem about formal systems.

THEOREM (Löwenheim-Skolem). Let M be a model for a collection T of constant and relation symbols. There is an elementary sub-model of M whose cardinality does not exceed that of T if T is infinite and is at most countable if T is finite.

The proof again uses the axiom of choice. Notice that in particular if M is a model for a certain set of statements, the sub-model will also be a model for these statements. Let N be an arbitrary subset of M containing all the constants $\bar{c}_\alpha$, and let $A(y,x_1,\ldots,x_n)$ be an arbitrary formula of n+1 variables. For each $\bar{x}_1,\ldots,\bar{x}_n$ in N, whenever there is a $\bar{y}$ in M such that $A(y,x_1,\ldots,x_n)$ is true in M at $\bar{y},\bar{x}_1,\ldots,\bar{x}_n$, choose one such and adjoin it to N. If we do this for all formulas A and all possible $\bar{x}_i$ we obtain a set N^* containing N. Since the number of n-tuples of elements of N is the same as the cardinality of N if N is infinite, countable if N is finite, and a similar statement holds concerning the number of formulas in the language T, it easily follows that the cardinality of N^* is at most that of $\bar{\bar{N}} + \bar{\bar{T}} + \aleph_0$ where we write $\bar{\bar{S}}$ for the cardinality of a set S and $\aleph_0$ is the cardinality of the integers. Define N_0 as the set of constants $\bar{c}_\alpha$, and put $N_{k+1} = N_k^*$. We now claim that if we define N' as the union of the N_k, then N' has the desired properties stated in the theorem. It is clear that N' satisfies the cardinality statement. Let $A(x_1,\ldots,x_n)$ be a formula with r quantifiers. The remainder of the proof proceeds by induction on r. If r is zero there is nothing to prove. Assume an induction hypothesis that for any formula $A(x_1,\ldots,x_m)$ with less than r quantifiers and any $\bar{x}_1,\ldots,\bar{x}_n$ in N', A is true in N' at $\bar{x}_1,\ldots,\bar{x}_m$ if and only if it is true in M at $\bar{x}_1,\ldots,\bar{x}_m$. We may clearly assume that A begins with a quantifier, and by perhaps replacing A by its negation, we can even assume that A is of the form $\exists\, yB(y,x_1,\ldots,x_n)$. Let $\bar{x}_1,\ldots,\bar{x}_n$ be arbitrary elements of N'. We now show that $A(x_1,\ldots,x_n)$ holds at $\bar{x}_1,\ldots,\bar{x}_n$ in M if and only if it holds there in N'. We know that all the $\bar{x}_i$ lie in N_k for some k. If there is a $\bar{y}$ in M such that $B(y,x_1,\ldots,x_n)$ is true in M at $\bar{y},\bar{x}_1,\ldots,\bar{x}_n$, there is also such a $\bar{y}$ in N_{k+1} and hence in N'. Since B involves r-1 quantifiers it then follows from our induction hypothesis that B is true at $(\bar{y},\bar{x}_1,\ldots,\bar{x}_n)$ in N'. This in turn means exactly that $A(\bar{x}_1,\ldots,\bar{x}_n)$ holds in N'. If there is no $\bar{y}$ such that B is true at $\bar{y},\bar{x}_1,\ldots,\bar{x}_n$ in M, then there can be no $\bar{y}$ such that it holds in N', since again by the induction hypothesis this would mean that it held in M. This completes the proof.

When the Löwenheim-Skolem theorem is applied to particular formal systems, we obtain as special cases: Every group, field, ordered field, etc., has a countable subsystem of the same type. A more spectacular result follows from applying the theorem to set theory (a system which we shall later formalize): There is a countable collection of sets, such that if we restrict the membership relation to these sets alone, they form a model for set theory (more precisely all the true statements of set theory are true in this model). In particular, within this model which we may denote by M, there must be an uncountable set. This paradox, that a countable model can contain an uncountable set, is explained by noting that to say that a set is uncountable merely asserts the nonexistence of a one-one mapping of the set with the set of integers. The "uncountable" set in M actually has only countably many members in M, but there is no one-one correspondence within M of this set with the set of integers. (Our discussion is necessarily incomplete since we have not discussed how to define integers in the formalism of set theory, nor whether the integers are necessarily in M.)

6. EXAMPLES OF FORMAL SYSTEMS

Let us state Peano's axioms in the usual form:

i) each integer has a unique successor

ii) there is an integer 0 which is not the successor of any integer

iii) two distinct integers cannot have the same successor

iv) if M is a set of integers such that 0 is in M, and such that if an integer x is in M then its successor is in M, then every integer is in M.

If one were to attempt a translation of these informal axioms into a formal system, one would most naturally use a constant symbol c_1 representing 0, and a binary relation symbol $S(x,y)$ representing "y is the successor of x". Axioms (i)-(iii) then have obvious translations into our formal system. Axiom (iv), on the other hand, admits no such exact translation. In describing a particular formal system, it is of course illegal to use words which cannot be translated into the formal system. The problem with axiom (iv) is the occurrence of the word "set". An examination of other informal axiom systems shows that the concept of set occurs frequently. For example, the definition of the reals as a complete ordered field speaks about an arbitrary bounded set of reals having a least upper bound. Because of such important examples, it might appear most reasonable to proceed directly to a formalization of the set concept. However, set theory brings with it many problems, such as the continuum hypothesis, which may be quite irrelevant to the particular subject at hand. In practice, the sets we need in a subject such as number theory, are those which can be described by fairly specific properties. Thus, an alternative approach is to avoid mentioning sets and to speak of certain properties instead. Since there are usually an infinite number of properties in which we are interested, we will in general be required to have infinitely many axioms. To be more specific, to formalize axiom (iv) above, one might introduce, for each formula A(x) with one free variable, an axiom of the form

$$(A(0) \ \& \ \forall x(A(x) \ \& \ S(x,y) \to A(y))) \to \forall x \ A(x)$$

As remarked in the last section, no such scheme can capture the full strength of axiom (iv), in the sense that one cannot in this way insure that the only model (up to isomorphism) of the axioms is the set of integers. However, using this form of the induction axiom is quite sufficient for many purposes. We shall now describe two formal systems which proceed in essentially this manner.

The axiom system we shall first discuss we shall call Z_1. It is a system which is very close to what might be called elementary arithmetic. The elements in the system are to be thought of as the non-negative integers. We introduce two ternary relations R_1 and R_2 corresponding to addition and multiplication. To make our formulas more readable we shall write $x + y = z$ and $x \cdot y = z$ in place of $R_1(x,y,z)$ and $R_2(x,y,z)$ respectively. We use two constant symbols, 0 and 1. Also, we introduce the notation $\exists! \, xA(x)$ as an abbreviation for $\exists x \, \forall y(A(y) \leftrightarrow x = y)$, (to formalize the notion "there exists a unique x such that A(x)").

<u>Axioms for</u> Z_1.

1. $\forall x, y \ \exists! \, z(x + y = z)$
2. $\forall x, y \ \exists! \, z(x \cdot y = z)$
3. $\forall x \ (x + 0 = x) \ \& \ (x \cdot 1 = x)$
4. $\forall x, y \ (x + (y + 1) = (x + y) + 1)$
5. $\forall x, y \ (x \cdot (y + 1) = x \cdot y + x)$
6. $\forall x, y \ (x + 1 = y + 1 \to x = y)$
7. $\forall x \ (\sim x + 1 = 0)$

Our eighth axiom consists actually of an infinite number of axioms and is more properly called an axiom scheme. To state it, we enumerate the countably many formulas of our system which contain at least one free variable, $A_m(x,t_1,\dots,t_k)$, where of course k depends on m. The particular enumeration we choose is of no consequence.

$$8_m. \ \forall t_1,\dots,t_k[(A_m(0,t_1,\dots,t_k) \ \& \ \forall y(A_m(y,t_1,\dots,t_k) \to A_m(y+1,t_1,\dots,t_k))) \to \forall x \ A_m(x,t_1,\dots,t_k)] \ .$$

Since we wish to be precise for the moment, it should be noted that we have used certain abbreviations in the statement of the axioms. In Axiom 4 we find the expressions $y + 1$ and $x + y$ occurring. This really means that we should restate Axion 4 as follows.

$$\forall x,y \ \exists u,v(u = x + 1 \ \& \ v = x + y \ \& \ x + u = v + 1)$$

Similarly Axiom 5 should be restated by introducing new letters u, v and w to stand for $y + 1$, $x \cdot y$ and $x \cdot y + x$. Since writing out our formulas in this detail would make our formulas too awkward, we shall often use such abbreviations.

The reader will observe that 1, 6 and 7 imply (i)-(iii) of Peano's informal axioms. Axiom 8 states the principle of mathematical induction, not for arbitrary sets, but only for those sets defined by an "arithmetical" condition A_m for some fixed $t_1,\ldots,t_k$. Since there are only countably many arithmetical conditions, and uncountably many sets of integers, it is clear that Axiom 8_m is weaker than the intuitive induction principle. What is indeed surprising is that although our system only allows the two relations $+$ and $\cdot$, it is sufficiently powerful to express all of what is traditionally called elementary number theory. The interest in presenting Z_1 in this form lies in the fact that its statements are essentially in the form of Diophantine equations. This will have implications about the theory of such equations and whether or not an effective method can be found for deciding if a given such equation can be solved. We postpone further discussion of Z_1 to a later section.

The second system we discuss is another formalization of elementary number theory, which we call Z_2. This system has the advantage that it is obvious that all of traditional elementary number theory can be formulated in it. In this system the elements are to be thought of as finite sets. The axioms will be very similar to the axioms we shall later give for set theory with the important exception that we shall not assume the existence of infinite sets. Certain sets will be defined to be integers and we shall state an induction axiom for these. As is well known, all the usual mathematical notions such as functions, relations etc., can be defined from the concept of set. This will mean that any argument concerning integers which speaks only about finite objects (although it may use induction to prove the existence of certain objects) can be formalized in Z_2. Again the induction axiom is an infinite scheme.

The only relation we use in Z_2 is a binary relation, ϵ, which denotes membership, so that $x \in y$ is to be read "x is a member of y". We use one constant symbol $\emptyset$, to denote the empty set.

Axioms for Z_2.

1. $\forall x,y\ (x = y \leftrightarrow \forall z(z \in x \leftrightarrow z \in y))$.
2. $\forall x\ (\sim x \in \emptyset)$
3. $\forall x,y\ \exists z\ \forall w\ (w \in z \leftrightarrow w = x \vee w = y)$
4. $\forall x,y\ \exists z\ \forall w\ (w \in z \leftrightarrow w \in x \vee w \in y)$

Before giving the axiom scheme for mathematical induction we make some observations. Axiom 1 is known as the axiom of extensionality and it asserts that a set is determined by its members. Thus in our system there are no "atomic" sets which contain no members other than the unique empty set. Axiom 2 states the property of the empty set. Axiom 3 says that given two sets x, y there is a set which contains precisely x and y as its members. This set is called the unordered pair of x and y and denoted by $\{x,y\}$. Axiom 4 asserts the existence of the union of x and y which is denoted by $x \cup y$. We write $\{x\}$ for $\{x,x\}$ and define the ordered pair $\langle x,y\rangle$ as $\{\{x\}, \{x,y\}\}$. It is a simple exercise to prove now that $\langle x,y\rangle = \langle u,v\rangle$ holds if and only if $x = u$ and $y = v$. The natural model for Axioms 1 through 4 is the model of all finite sets which can be built up from $\emptyset$ in a finite number of steps. More precisely, let S_n be the sequence of sets defined by setting $S_0 = \emptyset$, and S_{n+1} equal to the union of S_n and the set of all subsets of S_n. We then have $S_n \subseteq S_{n+1}$ and if we let M denote the union of the sets S_n, it follows that M satisfies our axioms. Indeed, it is quite simple to see, by induction, that every model for these axioms contains in a canonical fashion a submodel isomorphic to M. Of course, a model may contain infinite sets and even other elements which could not be intuitively thought of as sets. Our next problem is to decide how to define integers and the successor operation. The definition we use is the one given by von Neumann for an ordinal number which we shall later use in our discussion of set theory. Since in Z_2

we are describing only finite sets, the ordinals here are the usual integers. In M, the set α_n which we wish to denote the integers n is defined by setting $\alpha_0 = \emptyset$ and α_n equal to the set $\{\alpha_0, \ldots, \alpha_{n-1}\}$. The first few integers are thus: $\emptyset$, $\{\emptyset\}$, $\{\emptyset, \{\emptyset\}\}$, $\{\emptyset, \{\emptyset\}, \{\emptyset, \{\emptyset\}\}\}$ etc. Each integer is thus the set of preceding integers. Of course we cannot formalize this definition since we do not yet have induction. Observe however that $m < n \longleftrightarrow \alpha_m \in \alpha_n$. This motivates the following definition:

Definition. x is an integer if

1) $\forall y,z\ (y \in x \ \&\ z \in x \rightarrow y = z \vee y \in z \vee z \in y)$
2) $\forall y,z\ (y \in x \ \&\ z \in y \rightarrow z \in x)$

One easily sees that in our intuitive model M, the sets α_n are precisely the sets satisfying this definition. Namely, first observe that in M every non-empty set x contains a member y such that if $z \in x$ then $\sim z \in y$. That is, y is "minimal" with respect to the $\in$-relation. Now if x is an integer in M, a "minimal" element of x must be $\emptyset$ by 2). If y is the minimal element of $x - \emptyset$, then clearly by 1) and 2) y must be $\{\emptyset\}$. By repeating this operation, one can show that x must be equal to some α_n. Thus our definition of integer is a reasonable one.

Definition. If x is an integer, let $x + 1$ denote $x \cup \{x\}$.

We show that $x + 1$ is an integer. If y and z belong to $x+1$, either both belong to $\{x\}$, both belong to x, or one belongs to x and the other to $\{x\}$. If a set belongs to $\{x\}$ it must equal x, so that in the first case $y = z$, in the second since both belong to x and x is an integer 1) holds, and in the third we trivially have either $y \in z$ or $z \in y$. Thus 1) holds. If $y \in x + 1$ and $z \in y$, then if $y = x$, we have $z \in x$ so $z \in x + 1$. If $y \in x$ we again have $z \in x$, since x is an integer so $z \in x + 1$.

We can now state the induction axiom. Again $A_m(x, t_1, \ldots, t_k)$ range over all formulas with at least one free variable. We write 0 in place of $\emptyset$ and Int x as an abbreviation for the definition that x is an integer.

$$5_m.\quad \forall t_1, \ldots, t_k\ [(A_m(0, t_1, \ldots, t_k)\ \&\ \forall y\ (\text{Int } y\ \&\ A_m(y, t_1, \ldots, t_k) \rightarrow A_m(y + 1, t_1, \ldots, t_k))) \rightarrow \forall x(\text{Int } x \rightarrow A_m(x, t_1, \ldots, t_k))]$$

This axiom scheme is essentially identical to the one given for Z_1. Again, we have not intuitively captured the complete strength of the informal induction principle since not every possible set of integers, is described by a property A_m for suitable t_i. Since as has been already remarked, no set of axioms for the integers can be categorical, this is not an overpowering objection to our system. We now show how to develop conventional number theory in our system Z_2. First, we define a <u>function</u> as a set of ordered pairs such that a set occurs as a left member at most once. The domain of the function is the set of left members. Now, whenever a function is defined in elementary number theory it usually is defined as follows: there is a certain condition $B(x)$ which says that x is a set of ordered pairs defining a function on the integers $\leq n$ for some n, $x = \{\langle 1, f(1)\rangle, \ldots, \langle n, f(n)\rangle\}$ and that a certain recursion relation holds among the $f(k)$. As long as the condition B can be expressed in Z_2 one can then define the relation $\langle n, f(n)\rangle$ without using the infinite set which is the graph of f. For example, we can define $x + y = z$ by saying that there is a function f whose domain is the set of integers $\leq x$ (which is actually the integer $x + 1$) and such that $f(0) = y$, $f(k+1) = f(k) + 1$ for $k < x$, and $f(x) = z$. In this manner one can handle all definitions except those where the object being defined by induction is an infinite set. An example of a definition which cannot <u>a priori</u> be done in Z_2 is the following: Let us define the sequence of functions $f_n(x)$ by putting $f_0(x) = x + 1$, $f_n(0) = n + 1$, $f_{n+1}(x + 1) = f_n(f_{n+1}(x))$. Since we are defining a function, hence an infinite set, by induction our method does not immediately apply. Since however we only use finitely many values of f_n for each n the definition of $f_n(x)$ can be reworded in Z_2. An example of a function genuinely beyond the power of Z_2 would be difficult to give explicitly, although an obvious diagonal argument shows that such must exist.

For our final example of a formal system we turn to real-closed fields. Here the goal is to axiomatize, as well as possible, the real number system. One way to proceed is to axiomatize set theory and develop the real number system by the method of Dedekind cuts or Cauchy sequences. As explained before we would like to avoid getting involved in general set theory. The method we choose is analogous to the system

Z_1. Here the axioms we arrive at have been studied for their own sake and systems which satisfy them are known as real-closed fields. We use three relations, $x + y = z$, $x \cdot y = z$, and $x > y$ as well as constants 0 and 1.

Axioms for Real-Closed Fields

1. Field Axioms. These are the well-known axioms for a field which we do not bother to repeat here.

2. Order Axioms:

i) $\forall x,y\ (x = y \vee x < y \vee y < x)$

ii) $\forall x,y\ (\sim (x < y\ \&\ y < x))$

iii) $\forall x,y\ (x > 0\ \&\ y > 0 \rightarrow x + y > 0\ \&\ x \cdot y > 0)$

3_m. Completeness Axioms. Let $f(x)$ denote $a_0x^m + \cdots + a_m$. The axiom states

$$\forall a_0, \ldots, a_m, x, y(f(x) > 0\ \&\ f(y) < 0\ \&\ x < y \rightarrow \exists z(x < z < y\ \&\ f(z) = 0)).$$

The completeness of the real number system is thus only stated for a special case involving polynomials. However, one knows that these axioms determine the algebraic properties of the real numbers. Indeed, Tarski has proved that given any statement about real numbers which can be expressed in this formal system, either it or its negation is deducible from these axioms. Unfortunately most properties of real numbers which are investigated involve notions of continuity or notions concerning integers.

7. PRIMITIVE RECURSIVE FUNCTIONS

Our next object is to study the system Z_2 more closely. As remarked, all "normal" combinatorial arguments can be expressed in Z_2 since most proofs use the induction principle only as applied to finite sets. To make more precise statements it is convenient to consider the problem of determining which functions can be expressed in Z_2. In this section we define a class of functions, the primitive recursive functions, which are defined on the integers (or n-tuples of integers) and which take integral values. It should be emphasized that these functions are "real" mathematical objects and not objects of any formal system such as Z_1 or Z_2. Indeed, our main task will be to show that these functions can be represented within Z_1 and Z_2.

Definition. A function $f(n_1,\ldots,n_k)$ from integers to integers is called primitive recursive (p.r.) if it is constructed by means of the following rules:

1. $f \equiv c$ for some constant c is p.r.
2. $f(n_1,\ldots,n_k) = n_i$ for $1 \leq i \leq k$ is p.r.
3. $f(n) = n + 1$ is p.r.
4. If $f(n_1,\ldots,n_k)$ and $g_1,\ldots,g_k$ are p.r. then so is $f(g_1,\ldots,g_k)$.
5. If $f(0,n_2,\ldots,n_k)$ is p.r. and if $g(m,n_1,\ldots,n_k)$ is p.r. and we have $f(n+1, n_2,\ldots,n_k) = g(f(n,n_2,\ldots,n_k),n,n_2,\ldots,n_k)$ then f is p.r.

Most elementary functions of integers are p.r. For example, addition, multiplication, powers, factorial, the nth prime, are all p.r. The importance of the p.r. functions is that they are clearly effectively computable. That is, if we are given the definition of a p.r. function f using 1-5, and integers $n_1,\ldots,n_k$, we can (given enough time), compute $f(n_1,\ldots,n_k)$ by using the inductive scheme of the definition of f. One should not think, however, that the p.r. functions exhaust the class of effectively computable functions. Indeed, it is easy to see that one can effectively list all the schemes for generating p.r. functions of one variable, say $f_m(n)$, and in this way obtain an effectively computable function of two variables m and n. If we now put $g(n)=f_n(n)+1$, g cannot be a p.r. function, yet is certainly effectively computable. (In listing the p.r. functions of one variable one must actually first list all p.r. functions since Rule 2 allows us to form functions of one variable from functions of many variables.)

The p.r. functions should be thought of as the simplest class of computable functions. Once we have shown that these functions can be expressed in a system, then using the various logical operations available, we can express many non-constructive functions, such as the first exponent for which Fermat's last theorem is false if one exists, etc. It is fairly obvious that any p.r. function can be expressed in Z_2. We shall state this more precisely. First, for any integer n, say $n = 5$, even though Z_2 has no symbol 5, we shall use "5" as an abbreviation for $1 + 1 + 1 + 1 + 1$. (Actually using the latter is itself an abbreviation as mentioned before.)

THEOREM. If $f(n_1,\ldots,n_k)$ is a p.r. function, there is a formula $A(x_1,\ldots,x_k,y)$ in Z_2 such that the following holds:

i) $\forall x_1,\ldots,x_k \; \exists ! \, y \; A(x_1,\ldots,x_k,y)$ is true as a statement about integers and is indeed provable in Z_2.

ii) If $f(n_1,\ldots,n_k) = m$, then $A(n_1,\ldots,n_k,m)$ is provable in Z_2.

Comment: We remind the reader of the distinction between a formula being "true" (as a statement about integers) and being "provable" in Z_2. All statements about integers which are provable in Z_2 are certainly true in the integers. The converse is not true, as we shall see in Section 8. Also, the theorem which we really need in order to convince ourselves that Z_2 is a suitable language for informal number theory is actually more general. Namely one needs to know that the imbedding of p.r. functions is done in a "natural" way so that we can prove in Z_2 certain obvious properties. For example, if $f_1 = f_2(f_3)$, we can prove the statement which says about the corresponding relations A_1, A_2, A_3 that f_1 is the composite of f_2 and f_3. To state explicitly which properties we actually need is tedious and we shall not do this.

The proof of the theorem is essentially trivial. One merely has to show that the various ways of defining p.r. functions can be expressed in Z_2. The only interesting case is 5; we assume for simplicity that $k = 1$. Suppose then that $g(n_1,n_2)$ is p.r., and that f is defined by

$$f(0) = c$$

$$f(n + 1) = g(f(n),n)$$

The proof of the theorem is by induction on the complexity of the definition of f, and so we may assume that there is a formula $B(n_1,n_2,m)$ such that $g(n_1,n_2) = m$ if and only if $B(n_1,n_2,m)$ is provable in Z_2. Then $f(n) = m$ is represented by the formula

$$\exists f[(f \text{ is a function}) \ \& \ (\text{domain } f = \{x : x \leq n\}) \ \& \ (f(0) = c)$$
$$\& \ (f(n) = m) \ \& \ \forall x \ (x < n \rightarrow g(f(x),x) = f(x + 1))] \ .$$

It is clear that this can be written as a formula $A(n,m)$ in Z_2, and that this formula satisfies the theorem.

Our next goal is to prove an analogous theorem for the system Z_1. Notice that the above proof uses essentially the fact that in Z_2 we can speak about finite sequences of arbitrary length. Because it is not clear how to do this in Z_1, the proof becomes non-trivial. However, the interest in Z_1 is not so much as a system for formalizing conventional number theory. The system Z_2 is obviously more convenient for that purpose since we can directly speak about sets. Rather, it is that the statements of Z_1 are of a relatively simple type, using only $+$ and $\cdot$, and it is of interest to know that any proposition in Z_2 has an equivalent statement in Z_1. Thus when discussing the imbedding of functions in Z_1, we shall primarily discuss what statements of Z_1 are true in the integers and later point out what can be <u>proved</u> in Z_1. Since Z_1 has $+$ and $\cdot$, statements concerning divisibility, primes etc., can be stated in Z_1 and we shall not explicitly do this if it is fairly obvious.

LEMMA. There is a formula $B(d,i,x)$ in Z_1 for which the following hold:

i) $\forall d, i \ \exists ! \ x \ B(d,i,x)$. Thus x is a function of d and i which we write $x(d,i)$.

ii) $\forall n, M \ \exists d \ \forall i,j \ [i < j < n \rightarrow x(d,i) > M \ \& \ x(d,j) > M \ \& \ (x(d,i)$ and $x(d,j)$ are relatively prime$)]$.

Proof. In words, ii) says that for all n and M, we can find a d, such that $x(d,i)$ for $i < n$ are relatively prime and are greater than M. Observe that $x > y$ is an abbreviation for $\exists z(z \neq 0 \ \& \ x = y + z)$. Also "$x$ and y are relatively prime" can be written as

$$\forall u,v,z[x = u \cdot z \ \& \ y = v \cdot z \rightarrow z = 1] \ .$$

We now define $B(d,i,x)$ as $x = 1 + (i + 1)d$. If n and M are given, we can take $d = (\max(n,M))!$, so that clearly $x(d,i) > M$. Also, if $i < j < n$, then if a prime p divides $x(d,i)$ and $x(d,j)$ then p divides $(i-j)d$. By unique factorization, either p divides d or p divides $i-j$. If p divides d, then clearly p does not divide $x(d,i)$. If p divides $i-j$, then $p < n$, so that p divides d, which is impossible.

Let now $y(c,d,i)$ represent the remainder obtained when c is divided by $x(d,i)$. The formula $z = y(c,d,i)$ can be written in Z_1.

LEMMA. For any sequence $a_1,\ldots,a_n$ of integers, there exists c and d such that for $i \leq n$, $a_i = y(c,d,i)$.

Proof. Observe that as it stands the lemma itself cannot be stated in Z_1 since it speaks about finite sets of arbitrary length. We shall return to this later. To prove the lemma choose d so that $x(d,i)$ for $i \leq n$ is greater than $\max a_i$. By the Chinese remainder theorem we can now find the required c.

THEOREM. If f is a p.r. function, there is a formula A in Z_1 such that $f(n_1,\ldots,n_k) = x$ if and only if $A(n_1,\ldots,n_k,x)$.

Proof. Again it is clear that we need only consider case 5 in our definition of p.r. functions. Thus, assume that $f(0,n_2,\ldots,n_k)$ and g can be expressed in Z_1. Let $A(n_1,\ldots,n_k,x)$ be the formula

$$\exists c,d\{f(0,n_2,\ldots,n_k) = y(c,d,0) \;\&\; x = y(c,d,n_1) \;\&$$
$$\forall i[i < n_1 \rightarrow y(c,d,i+1) = g(y(c,d,i),i,n_2,\ldots,n_k)]\}$$

In words, this formula means that for some c, d, the sequence $a_i = y(c,d,i)$ is such that $a_0 = f(0,n_2,\ldots,n_k)$ and $a_{i+1} = g(a_i,i,n_2,\ldots,n_k)$ and $a_{n_1} = x$. Since any sequence $a_0,\ldots,a_{n_1}$ is of the form $y(c,d,i)$ for some c and d, this is exactly the definition of f and the theorem is proved.

We turn to the question of provability in Z_1. This question has interest in its own right, but is not necessary for our main application. Our first lemma used the following facts:

i) if two numbers a and b have a common divisor, then they have a prime as a common divisor;

ii) if a prime p divides ab either p divides a or p divides b

iii) for each N, there is a number (in the proof taken as $N!$) which is divisible by all $n \leq N$.

First note that the induction axiom in Z_1 allows us to say that for any property $A(n)$ in Z_1, there is a least n such that $A(n)$. To prove i) we observe that we can thus prove in Z_1 the existence of a least non-trivial common divisor which is clearly prime. To prove ii) requires observing that the usual proof can be formalized in Z_1. Namely, we know there is a least positive d of the form $mp + na$, where m, n are positive or negative integers. (It is easy to avoid speaking about negative integers, which we do not have in Z_1.) As in the usual proof, if p does not divide a, then $d = 1$. The equation $b = mpb + nab$ now shows p divides b. We prove iii) by induction. Namely, if x is divisible by all $n \leq N$, clearly $x(N+1)$ is divisible by all $n \leq N+1$.

The second lemma can be restated as follows:

LEMMA. For all c, d, n, a if $x(d,i)$ are relatively prime for $i \leq n + 1$, then there is a c' such that $y(c,d,i) = y(c',d,i)$ for $i \leq n$, and $y(c,d,n+1) = a$.

This statement has the same intuitive content as the other formulation, since it allows us to extend sequences by one element at a time. To prove it, one first proves by induction on i, that for all $i \leq n$, there exists a t such that $x(d,n+1)$ is prime to t, and for $j \leq i$, $x(d,j)$ divides t. Thus there exists a t such that $x(d,i)$ divides t for all $i \leq n$ and $x(d,n+1)$ is prime to t. It follows easily that for some u, if we put $c' = c + tu$, c' has the desired property.

We can now prove in Z_1 that the formula given to define the p.r. function f, does indeed define a unique function. The proof is a simple exercise in induction which we leave to the reader. Also all the desired obvious properties of p.r. functions, such as composition, etc., can be proved in Z_1.

The fact that $+$ and $\cdot$ suffice to define p.r. functions is rather remarkable. Using it we can show that the system Z_2 can be "imbedded" in Z_1 and hence that Z_1 is as powerful as Z_2. More precisely we prove that every theorem in Z_2 about integers can be proved in Z_1. To see this, define a sequence of finite sets S_n and a sequence of integers n_k such that $n_0 = 0$, $S_0 = \emptyset$ and such that for each k, $n_k < n_{k+1}$ and the sets S_i for $n_k < i \leq n_{k+1}$ enumerate in a definite fashion all sets whose members are chosen from among the S_j with $j \leq n_k$. It is easy to see that this can be done so that the function $f(m,n)$ defined as 1 or 0 according as S_m does or does not belong to S_n is a p.r. function. One can now prove the axioms of Z_2 in Z_1 if we replace the relation $x \in y$ by $f(x,y) = 1$. In this way Z_2 is imbedded in Z_1 and the stated result can be shown to follow. The fact that any statement provable in Z_1 is provable in Z_2 follows immediately from the fact that $+$ and $\cdot$ are p.r. functions and so can be defined in Z_2.

8. GENERAL RECURSIVE FUNCTIONS

In certain mathematical situations, the question arises as to whether a particular function is "effectively computable". If an obviously effective procedure can be given for computing the functions, then the question will be considered answered. If, however, one feels that there is no such effective procedure, there is no hope of proving this until a precise definition of effectively computable functions has been given. As remarked in the previous section, any such definition should properly include the class of primitive recursive functions. Several such definitions have been given, which happily all give rise to the same class of functions, the general recursive functions.

One of the motivations for considering this whole question is the so-called "decision problem". Consider the formal system Z_1 (or equivalently Z_2). The various statements of Z_1 can be enumerated in a sequence A_n. This sequence of statements includes all questions of number theory currently being considered, as well as most questions in algebra and topology after suitable reformulation. Having solved the problem of giving a mechanical procedure for listing all these problems (which of course was historically a great step), there arises the possibility of mechanically deciding which are true. Thus, if we define $\varphi(n)$ as 1 or 0 according as A_n is true or false we can ask if $\varphi(n)$ is general recursive. If so, we have an effective means of deciding any question in Z_1. (Of course, showing that a given computation actually computes $\varphi(n)$

is the main difficulty, but here we are merely concerned with the exist-ence of such a computation). We shall show in this section that $\varphi(n)$ is not general recursive, a result which we can paraphrase by saying that there is no effective method for even "guessing" the truth or falsity of every statement in Z_1. A fortiori, using any consistent set of axioms, there is no effective method which given a statement A in Z_1, yields a proof of either A or $\sim A$. Thus the so-called decision problem cannot be solved. (It has been said that a famous mathematician once briefly believed he had found a decision procedure, but luckily for the rest of us, he was mistaken.)

The first definition of general recursive function was given by Gödel in 1934 who built on a suggestion of Herbrand. Before giving this definition it should be pointed out that there might appear, at first glance, an insuperable obstacle to giving such a definition. Namely, if one had an "effective" definition of recursive functions since there are only countably many instructions one can give for performing a calculation, one could enumerate these functions (or at least the functions of one variable), $f_n(x)$, and then put $g(n) = f_n(n) + 1$. The function g would have to be considered recursively defined, yet would not belong to the class of functions. The way out of this dilemma is to give a non-effective definition of a recursive function! This means that we state a criterion for a function to be recursive, yet we cannot effectively enumerate them. Gödel's definition proceeds from the observation that a p.r. function, for example, is completely described by a finite number of equations involving functions. Of course, for p.r. functions these equations must be of a very special type. In general, a given set of equations will not determine a function. Gödel's definition is essentially that a function is general recursive if there is some finite set of equations involving the function and auxiliary functions, which define the function uniquely.

To proceed further, it is necessary to develop a formal theory of calculations of functions. Our alphabet will consist of function symbols f, g, h,... each associated with a fixed number of variables, integer variables x, y, z,..., the symbols 0, =, and ', where x' represents x + 1, and finally parentheses (,) and the comma , . We make the following definitions:

1. A numeral is an expression on the form 0, 0', 0", ... etc.
2. A term is defined by
   (a) 0 is a term
   (b) the variables $x, y, z, \ldots$ are terms.
   (c) If r is a term, so is r'.
   (d) If $r_1,\ldots,r_n$ are terms, f a function symbol of n variables, then $f(r_1,\ldots,r_n)$ is a term.
3. An equation is an expression of the form $r = s$, where r and s are terms.
4. If E is a finite set of equations, an equation is a deduction from E if it can be obtained by repeated applications of the following rules.

Rule 1.

Given an equation we may replace all occurrences of a given variable x by a given numeral.

Rule 2.

If $f(n_1,\ldots,n_k) = m$ has been deduced, where m and n_i are numerals, then given an equation we may replace any occurrence of $f(n_1,\ldots,n_k)$ by m.

Rule 3.

If $r = s$ is deduced, then so is $s = r$.

Definition. A function $f(x_1,\ldots,x_k)$ is general recursive (or sometimes just "recursive"), if there is a finite set of equations such that for any choice of the numerals $n_1,\ldots,n_k$, there is a unique m such that $f(n_1,\ldots,n_k) = m$ can be deduced.

We give some examples. The system

$$f(x,0) = x \ ; \qquad f(x,y') = f(x,y)'$$

defines $f \equiv x + y$. This is a special case of the fact, whose proof is left to the reader, that every p.r. function is general recursive. The equations

$$f(0,m') = m'$$
$$f(n,0) = n'$$
$$f(n',m') = f(n,f(n',m))$$

define a function $f(x,y)$ for which it can be shown that if $g(y)$ is p.r., then for some x, $\forall y\ f(x,y) > g(y)$. In general, it is not at all obvious which equations will define functions. Thus, our second example clearly defines f by a more complicated "double" recursion than primitive and still more complicated examples can be given.

The statement that the general recursive functions, as we have defind them, exhaust the class of "effectively computable" functions is known as <u>Church's thesis</u>. Being essentially a philosophical statement there is no question of proof or disproof. However, there is strong evidence for it, principally the fact that all known computable functions are general recursive, and that quite different attempts to define general recursive functions have led to the same result. One of these attempts is due to Turing.

A <u>Turing machine</u> is a device through which an infinite tape is fed, the tape being divided into an infinite sequence of boxes. At each moment a given box is being "scanned" and the machine is in one of finitely many "states" $S_1, S_2, \ldots, S_n$. Each box is either empty or contains the symbol 1. The machine, after scanning a box does the following:

1. It either prints the symbol 1 in that box if it is empty, erases the symbol if it already appears, or leaves the box unchanged.

2. It moves the tape one box to the left or right.

3. It goes into a new state.

Thus to describe 1. we need a function $\varphi_1(i,j)$ defined for $0 \leq i \leq 1$, $1 \leq j \leq n$, and $0 \leq \varphi_1 \leq 1$ so that if the machine is in S_j and no symbol appears $\varphi_1(0,j)$ is 0 if the box is left unchanged, 1 if it is changed, and similarly for $\varphi_1(1,j)$ if the box has a symbol in it. To describe 2, we need a function $\varphi_2(i,j)$ of the same type to determine if the machine shifts left or right. To describe 3, we need a function $\varphi_3(i,j)$ where $1 \leq \varphi_3 \leq n$. Thus the machine is completely described by these three functions. To calculate, we prepare the tape so that m consecutive boxes are occupied, have the occupied box to the extreme left scanned and put the machine in S_1. If the machine reaches a state, say S_n, which it never leaves and never shifts or alters the tape, we say the calculation has ended provided the tape is blank except for a consecutive string, $f(m)$, of occupied boxes and the extreme left box of this string is being scanned. If this is true for all m, we say the machine calculates $f(m)$. One sees how this machine attempts to duplicate the way one actually performs computations. It can now be shown that the class of functions which are calculated by Turing machines coincides with the class of general recursive functions ([15]). We do not do this here, but remark that it is not at all surprising that the rules governing the operation of a Turing machine can be written in the form of equations connecting what appears on the tape and the state of the machine. Conversely, if we have a set of equations it is a tedious exercise in programming to design a machine to search through all deductions. In our approach we think of recursive functions as defined by equations. Although an occasional argument uses the words Turing machine, it should be clear how to achieve the same effect with suitable equations.

If S is a set of integers, we shall let χ_S denote the characteristic function of S, i.e., $\chi_S(n) = 0$ or 1 according as $\sim n \in S$ or $n \in S$.

Definition. A set S is recursive if its characteristic function is general recursive.

A set S is recursively enumerable if S is empty, or if S is the range of a general recursive function.

The question of computing χ_S, where S is the range of a given general recursive function occurs frequently in mathematics. For example, for a fixed k let us enumerate all pairs consisting of a polynomial $p(x_1,\ldots,x_k)$ with integral coefficients and a k-tuple $(n_1,\ldots,n_k)$ where n_i are integers. Define a function on this sequence taking as its value the polynomial $p(x_1,\ldots,x_k)$ if $p(n_1,\ldots,n_k) = 0$, otherwise let it take as its value the zero polynomial. The range of this function is the set of all p such that $p(x_1,\ldots,x_k) = 0$ has an integral solution. Hilbert's tenth problem is precisely to determine the range of this function or in other words, given a polynomial to determine effectively if it has an integral solution. It is not known whether such a suitable general recursive function exists. Other examples, such as the word problem for groups and semi-groups have also been considered and in these cases it has been shown that general recursive functions having certain desired properties do not exist. Here we shall content ourselves with showing that there are "unsolvable problems" in this sense. More precisely we have the following theorem which is the fundamental result in this subject.

THEOREM. There exists a recursively enumerable set which is not recursive.

Our proof will specifically exhibit such a set. Let us first enumerate in any "effective" manner, all possible finite sets of equations in which a function symbol f of one variable occurs. Now enumerate all possible deductions from these equations. Since these deductions are countable such an enumeration is possible. If the nth system of equations defines a function, we denote this function by f_n. Even if f_n is not defined, we shall write $f_n(a) = b$ to mean that there is a deduction from the n-th set of equations of the form $f(a) = b$, where a and b are integers. We now enumerate all deductions of the form $f_n(a) = b$. It can easily be shown that all our enumerations can be so chosen that there exist p.r. functions $\varphi_1, \varphi_2, \varphi_3$ such that if the equation $f_n(a) = b$ occurs in the m-th place of our sequence, then $n = \varphi_1(m)$, $a = \varphi_2(m)$, $b = \varphi_3(m)$. Let g, h_1, h_2 be p.r. functions such that the map $(a,b) \to g(a,b)$ is a 1-1 correspondence between the set of all pairs of integers and the set of integers, and such that $a = h_1(g(a,b))$, $b = h_2(g(a,b))$. We now define $F(n)$ as follows: Let $r = h_1(n)$ and $s = h_2(n)$. If there is a deduction of the equation $f_r(r+1) = 0$ occurring in our list before the s-th place, put $F(n) = r+1$. If not, put $F(n) = 0$. Our first claim is that F is p.r. This can be checked by explicitly writing out all the maps used. It should be fairly obvious that all the functions we have used were defined by quite simple recursion procedures. Observe that for all r, $r+1$ occurs in the range if and only if $f_r(r+1) = 0$ can be deduced. Let S be the range of F, χ its characteristic function. We claim that for no r can f_r be equal to χ. (If f_r is not defined this statement is vacuous.) For, if f_r is defined then $f_r(r+1)$ must be eventually computed. If $f_r(r+1) = 0$ then $r+1$ is in the range so $\chi(r+1) = 1$. If $f_r(r+1) \neq 0$, then $r+1$ is not in the range and $\chi(r+1) = 0$. In either case $\chi(r+1) \neq f_r(r+1)$ and the theorem is proved.

COROLLARY. The decision problem for Z_1 (or Z_2) is unsolvable.

<u>Proof</u>. The statement $n \in S$, (or equivalently $\exists\, m(f(m) = n)$), for each n is a statement in Z_1 and we have seen that no general

recursive function can tell us the truth or falsity of even this limited class of statements.

Let us recall that the statements of Z_1 can all be put in the form $Q_1x_1,\ldots,Q_nx_nA(x_1,\ldots,x_n)$, where Q_i are quantifiers and A is a propositional function of equations of the form $p(x_1,\ldots,x_n) = 0$ where p is a polynomial with integral (perhaps negative) coefficients. By using elementary number-theoretic devices it can easily be shown that one can even assume that A is of the form $p(x_1,\ldots,x_n) = 0$. (One uses the facts

$$y_1 = 0 \mathbin{\&} \cdots \mathbin{\&} y_r = 0 \leftrightarrow y_1^2 + \cdots + y_r^2 = 0 ;$$

$$y \neq 0 \leftrightarrow \exists z_1,\ldots,z_4(z_1^2 + \cdots + z_4^2 + 1 = y^2) ,$$

and

$$y_1 = 0 \vee \cdots \vee y_r = 0 \leftrightarrow y_1y_2 \cdots y_r = 0.)$$

Thus Diophantine equations, where we allow both universal and existential quantifiers, do not allow a decision procedure. Further work has been done to show that if one allows variable exponents the resulting class of "Diophantine equations" with only existential quantifiers also does not allow a decision procedure.

In the above proof, the recursively enumerable set S is actually the range of a p.r. function. This is however always the case.

THEOREM. Every non-void recursively enumerable set is the range of a p.r. function.

Proof. Let S be the range of a general recursive function f. We may clearly assume S is not empty and assume $a \in S$. The function f is defined by some system of equations. We assume all the deductions of this system have been enumerated. There is then a p.r. function g, such that if the m-th deduction is of the form $f(x) = y$, then $g(m) = y$, otherwise $g(m) = a$. The range of g is then S.

One can even show that the general recursive functions can be obtained from the p.r. functions by allowing just one additional operation. If $f(y,x_1,\ldots,x_n)$ is a function let $\mu_y f(y,x_1,\ldots,x_n)$ denote the function $g(x_1,\ldots,x_n)$ defined by

i) $g(x_1,\ldots,x_n) \equiv 0$ if for any $\bar{x}_1,\ldots,\bar{x}_n$ $\forall y\, f(y,\bar{x}_1,\ldots,\bar{x}_n) \neq 0$.

ii) If i) does not apply then $g(x_1,\ldots,x_n) = a$ if a is the <u>least</u> y such that $f(y,x_1,\ldots,x_n) = 0$.

THEOREM. The general recursive functions form the smallest class of functions containing the p.r. functions and which is closed under composition and the μ-operator.

<u>Proof</u>. If $f(y,x_1,\ldots,x_n)$ is recursive and for all $x_1,\ldots,x_n$ $\exists y\ f(y,x_1,\ldots,x_n) = 0$, then we can define a Turing machine to calculate $\mu_y f(y,x_1,\ldots,x_n)$ by merely allowing the Turing machine for f to calculate with successive values of y until a value is found such that $f(y,x_1,\ldots,x_n) = 0$. Since the process must terminate, $\mu_y f$ is general recursive. Thus the general recursive functions are closed under the μ-operator. On the other hand if $h(x)$ is a general recursive function, there is clearly a p.r. function $f(x,y)$ such that $f(x,y) = 0$ if and only if allowing the Turing machine for h to calculate for y steps with the argument x succeeds in calculating $h(x)$. Put $g(x)=\mu_y f(x,y)$. There is also a p.r. function $k(x,y)$ which gives the value on the tape, (if any) after allowing the machine to calculate y steps with the argument x. Clearly $h(x) = k(x,g(x))$, so the theorem is proved.

9. GÖDEL INCOMPLETENESS THEOREM

The theorems of the previous section are not results about what can be proved in particular axiom systems; they are absolute statements about functions. Historically, it was Church who first showed the existence of "unsolvable" problems. However, there is an older result due to Gödel which asserts the existence of propositions which are undecidable in particular formal systems. The two types of results are closely related and are both proved by similar diagonalization methods. We shall derive Gödel's theorem from our previous results and later give the original method.

<u>Gödel Incompleteness Theorem</u>

There is a statement A in Z_1 such that neither A nor $\sim A$ can by proved from the axioms of Z_1.

Our proof explicitly gives A and will incidentally show that A is false when interpreted in the natural model of integers. The theorem holds equally well for Z_2 and indeed for more general systems, as will be shown.

The first step is the so-called <u>arithmetization</u> of Z_1. By this we mean that we assume the formulas of Z_1 are enumerated in a natural manner. (For example, a suitable lexicographic ordering based on the length of the formula.) Once this is done, the various elementary operations such as forming negation and conjunction of formulas, bounding a free variable, etc. become rather simple p.r. functions on the integers corresponding to the formulas. These integers are usually called the Gödel numbers of the formulas. (There is, of course, no canonical fashion of attaching integers to formulas.) Now, the rules of the predicate calculus are given quite effectively. Also, although there are infinitely many axioms of Z_1, they are generated by a simple rule and one can easily enumerate their Gödel numbers. This means that by enumerating all possible proofs, we see that the <u>provable</u> statements (or rather their Gödel numbers) are the range of a p.r. function.

In § 8, we explicitly constructed a p.r. function F whose range is not recursive. More precisely, we know that if f_r is the recursive function defined by the rth set of equations, then r+1 is in the range of $F \longleftrightarrow f_r(r+1) = 0$. The function $F(x)$ can be represented in Z_1 in the sense that there is a formula in Z_1 which says $F(x) = y$. Let B_n denote the statement "n is in the range of F". The Gödel numbers of B_n and of $\sim B_n$ are clearly p.r. functions of n. We now consider the following Turing machine (or an equivalent set of equations defining a general recursive function): For each n, we go through the list of provable statements in order until we encounter a proof of B_n or of $\sim B_n$. If we do, the calculation ceases and we put $g(n) = 1$ in the first case, $g(n) = 0$ in the second. If neither B_n or $\sim B_n$ is provable then $g(n)$ remains undefined. (Observe that we have not a priori excluded the possibility that both B_n and $\sim B_n$ are provable. In that case $g(n)$ would depend upon which proof occurs first.) The function g which has been very explicitly defined must be f_r for a particular r which can be explicitly found. Let A be the statement "r+1 is in the range of F". By the characteristic property of F that $f_r(r + 1) \neq \chi(r + 1)$, where χ is the characteristic function of the range of F, the following holds:

Assertion

If a proof of A occurs before a proof of $\sim A$, then $r+1$ is not in the range of A.

If a proof of $\sim A$ occurs before a proof of A, then $r+1$ is in the range of A.

Now a statement which is provable in Z_1 is of course true in the model of integers. Also since Z_1 is consistent not both A and $\sim A$ are provable. Thus, if A is provable, the assertion implies $r+1$ is not in the range of F. But if A is provable, then A is true and $r+1$ is in the range, a contradiction. If $\sim A$ is provable, then by the assertion $r+1$ is in the range. But if $\sim A$ is provable, then $\sim A$ is true and $r+1$ is not in the range, again a contradiction. Thus neither A nor $\sim A$ is provable and the theorem is proved.

COROLLARY. A is false in the integers.

Proof. If $r+1$ were in the range of F_1 then for some m, $F(m) = r+1$. Since F is recursive, we can write down all the steps of the calculation of $F(m)$. These calculations would then constitute a proof in Z_1 of $F(m) = r+1$ and hence a proof of A. But A is unprovable in Z_1 so the corollary is proved.

We have now arrived at a rather peculiar situation. On the one hand $\sim A$ is not provable in Z_1 and yet we have just given an informal proof that $\sim A$ is true. (There is no contradiction here since we have merely shown that the proofs in Z_1 do not exhaust the set of all acceptable arguments.) This suggests that a close analysis of the proof of the corollary will lead us to a natural principle which was used in the proof and which cannot be proved in Z_1. (The statement A itself is certainly not a natural statement but was merely designed to be undecidable.)

To state the theorem which emerges, we must recall that by our assignment of Gödel numbers to formulas, we have translated statements about the formal system Z_1 into statements about integers. Thus in particular the statement that Z_1 is consistent, or equivalently, that $0 = 1$ does not occur in the list of provable statements, now becomes the fact that a certain integer is not in the range of a certain p.r. function. Since the p.r. functions are representable in Z_1, this is thus a single statement in Z_1 which we denote by Consis Z_1.

Gödel's Second Incompleteness Theorem.

Consis Z_1 cannot be proved in Z_1.

Proof. Assume that Consis Z_1 can be proved in Z_1. We now give an argument about the formal system Z_1 which is presented informally, but which can be completely transcribed in Z_1 by means of the Gödel numbering. (This is perhaps the most subtle point in the entire subject and the reader is urged to stop and review carefully what we have done up to here.)

We first repeat the proof of our Assertion. This can be easily transcribed in Z_1 since we only used simple finitistic arguments. We did, of course, refer to recursive functions in the course of the proof but this can be replaced by enumerating all possible defining equations and speaking about the corresponding numbers. We also can enumerate all possible deductions and thus speak about the values of the functions all completely within Z_1. We have not used any special property of the recursive function g. (In particular, we did not use the fact that Z_1 is consistent.) Now, since we have assumed that Consis Z_1 is provable we proceed as follows. If r+1 is in the range of F, say, F(m) = r+1, then the computation would be a proof of A. Since Z_1 is consistent, the first alternative of the Assertion holds, r+1 is not in the range which is a contradiction. Thus r+1 is not in the range, or $\sim A$ is true.

We have thus given a proof in Z_1 of $\sim A$. This is not possible. Hence Consis Z_1 is unprovable in Z_1 since that would make $\sim A$ provable. This proves the theorem.

We now briefly sketch the original proof of the First Incompleteness Theorem. This proof is actually shorter and although it uses essentially the same argument, it does not use the notion of a recursive function. (Our only reason for proceeding as we did was to make the proof appear more natural in the context of recursive functions.) One uses a diagonalization argument directly on the statements of Z_1. First, enumerate in a natural way all formulas with one free variable $A_n(x)$. Let $B(n)$ be the statement that $A_n(n)$ is not provable. Then $B(n)$ must be $A_{n_0}(n)$ for an integer n_0 which can be explicitly computed. The statement $B(n_0)$ then intuitively says that it itself is not provable.

More precisely, if $B(n_o)$ is provable then it is true and so $A_{n_o}(n_o)$ is not provable. But, $A_{n_o}(n_o) \equiv B(n_o)$ so $B(n_o)$ is not provable which is a contradiction. If $\sim B(n_o)$ is provable, then $B(n_o)$ is false and hence $A_{n_o}(n_o) \equiv B(n_o)$ is provable, again a contradiction. Thus neither $B(n_o)$ nor $\sim B(n_o)$ is provable.

Remarks. Although it is of secondary interest, the Second Incompleteness theorem can itself be transcribed as a statement in Z_1. One can ask what principles were used, other than the axioms of Z_1, in its proof. The original proof of Gödel used, as he pointed out, a more powerful statement than the consistency of Z_1.

Definition. A set S of statements in Z_1 is said to be ω-consistent if it is impossible to deduce from S the statement $\exists\, x\, B(x)$, and for each numeral n to also deduce $\sim B(n)$.

Clearly, if the statements of S are true in the integers, then S is ω-consistent. On the other hand, consider the set S consisting of the axioms of Z_1 and the statement $\sim$ Consis Z_1. Then, S is consistent by the Second Incompleteness Theorem. (Of course, $\sim$ Consis Z_1 is false in the integers.) Now, $\sim$ Consis Z_1 says $\exists\, n$ such that n is the Gödel number of a proof of a contradiction. Yet for each n, we can examine the nth proof and see that it does not yield a contradiction. Thus S is not ω-consistent. Intuitively, we can say that the notion of consistency is only a crude approximation to the notion of truth, while ω-consistency is a slightly better approximation.

Now Gödel observed that his proof used the ω-consistency of the axioms of Z_1. Rosser later showed that by slightly modifying the argument one need only assume the consistency of Z_1. Thus, for example, if S consists of the axioms of Z_1 with $\sim$ Consis Z_1, the proof can be repeated to show that Consis S cannot be proved in S. Admittedly, S is not a natural system since it contains false statements.

An important point which lies behind the Second Incompleteness Theorem is the fact that the notion of truth cannot be formalized in Z_1. This is a result of Tarski which more precisely says the following: there is no formula A(n) in Z_1 such that for each numeral A(n) is true in the integers if and only if the nth statement is true in the integers.

The proof resembles that of the Incompleteness Theorem except that we deal with truth rather than provability. Assume that such a formula $A(n)$ exists. Let $B_n(x)$ denote a natural enumeration of formulas with one free variable. The statement $B_n(n)$ will then be the $f(n)$-th statement where $f(n)$ is a simple p.r. function. Our hypothesis is then $A(f(n)) \leftrightarrow B_n(n)$. Now the statement $\sim A(f(n))$ must be of the form $B_{n_o}(n)$ for some n_o. Thus we have $B_{n_o}(n) \leftrightarrow \sim B_n(n)$. Now put $n = n_o$ so that $B_{n_o}(n_o) \leftrightarrow \sim B_{n_o}(n_o)$ which is clearly absurd and the theorem is proved. Again our proof is explicit and yields for each $A(n)$ a specific statement, $\sim B_{n_o}(n_o)$ for which the "truth function" breaks down.

We can also see that if the notion of truth were formalizable by such a property $A(n)$ and the usual properties of truth could be proved to hold in Z_1, then this would yield a proof of Consis Z_1 in Z_1. For, it would be simple to show that all proved statements are true, and that a statement and its negation cannot both be true. We leave as an exercise for the reader the proof of the following fact: For each r, there is a formula $A(n)$ in Z_1 such that if we enumerate all statements T_n in Z_1, which have fewer than r quantifiers, in a natural way, then $T_n \leftrightarrow A(n)$ is true for all n.

Lastly, we close with some philosophical remarks. Gödel's theorems destroyed the hopes of Hilbert's program of proving the consistency of higher systems by finitistic methods which would be acceptable to everyone. Indeed the program failed already for Z_1. Nevertheless, work has been done to give consistency proofs which use as few "higher" principles as possible. Perhaps the most interesting result is due to Gentzen. His point of view is essentially that one can introduce "higher" principles by using transfinite induction up to higher ordinals. If we define $\omega_1 = \omega$, $\omega_{n+1} = \omega^{\omega_n}$, and finally $\epsilon_o = \lim_{n \to \infty} \omega_n$, then ϵ_o is a countable ordinal. Gentzen gave a proof of Consis Z_1 by defining a well ordering of the proofs in Z_1 or order type ϵ_o and then showing that for every proof of a contradiction, there was another such proof which precedes it. Thus if induction up to ϵ_o is accepted as an axiom, Consis Z_1 can be proved. (It is true that the well-ordering of integers corresponding to ϵ_o can be expressed by a formula in Z_1. Gentzen's work shows that we cannot prove that for every property P(n) in Z_1, there is a least element with respect to the ordering ϵ_0 and satisfying P(n).) Furthermore, he showed that any ordinal less than ϵ_0 can be proved to be a well-ordering (in a suitable sense), within Z_1. Thus if one measures the strength of a system by the ordinals which can be handled in the system, ϵ_0 is the precise strength of Z_1.

10. GENERALIZED INCOMPLETENESS THEOREM

As stated, the Incompleteness Theorems dealt a death-blow to the Hilbert program. The possibility remains, however, that if we suitably extend our notion of proof we can prove the consistency of various systems. Indeed by allowing more complicated systems than Z_1 one can prove Consis Z_1. However, if one wishes to give a fixed system Σ in which one can prove the consistency of any mathematical system, then this cannot be done. More precisely,

GENERALIZED INCOMPLETENESS THEOREM. Let Σ be a formal system whose axioms are given by some recursive rule. If Σ is consistent, and the p.r. functions can be imbedded in Σ, then Consis Σ cannot be proved in Σ.

Strictly speaking, we have not stated a precise mathematical theorem since we have not stated exactly what is meant by imbedding the p.r. functions. In practice there is no difficulty in seeing that the theorem applies to all the usual systems considered. The proof of the theorem consists merely of observing that no particular properties of Z_1 other than those mentioned in the theorem were used in the proof of the original incompleteness theorems. (We actually did use the fact that proved statements were true in the integers. As remarked, a careful analysis shows that only ω-consistency and even only consistency is necessary to get around this step. We avoid these fine points and the reader can content himself with the fact that we have certainly given a full proof in the case in which the axioms of Σ are true in the intuitive model.) The requirement that the axioms be given recursively is essential; otherwise we could take for Σ the set of all true statements of Z_1.

There is an important related aspect of the incompleteness theorem which has great philosophical consequence. Namely, if we consider the statement Consis Z_1 or Consis Σ for higher systems such as set theory, these are elementary statements in Z_1 which are true and yet cannot be proved without appealing to our intuition about higher systems. This means that Consis ZF, where ZF is the system for set theory we shall later study, cannot be proved except by reasoning about infinite sets of the type of $\aleph_1$, $\aleph_\omega$, etc., since all these can be described in ZF. Thus if one dismisses these higher infinites as not having any true significance for mathematics, one also gives up all hope of deciding certain number theoretic statements of the most elementary kind. In our opinion this is the gravest weakness of the Intuitionist position since statements which they admit as meaningful have no hope of ever being decided according to their requirements.

11. FURTHER RESULTS IN RECURSIVE FUNCTIONS

The theory of recursive functions has also been extensively developed for its own sake. This is due in part to the general interest in computing machines which has recently developed. Here we shall touch on a few points.

Definition. We say that a function f is defined recursively in g, if there is a finite set of equations E involving the function symbols f and g (and possibly other function symbols as well), such that if we adjoin the infinite table of the values of g, $g(1) = n_1$, $g(2) = n_2, \ldots,$ each value of f can be uniquely deduced.

Intuitively, this means that f can be computed by a recursive process which may, at any time, pause and ask for the value of g at some number n. We have already shown the existence of a p.r. function F, such that if χ is the characteristic function of its range, and if f_r is the recursive function (if any) defined by the rth set of equations, then $\chi(r+1) = 0$ unless $f_r(r+1) = 0$ in which case $\chi(r+1) = 1$. We shall now show that this non-recursive χ is indeed "maximally" non-recursive.

THEOREM. Let g be a recursive function, χ_g the characteristic function of its range. Then χ_g can be recursively defined in χ.

Proof. Given n, we shall show how to find $\chi(n)$. Consider the set of equations corresponding to the (attempted) calculation of the following function H. For each k, we enumerate the range of g searching for n. If we encounter n we put $H(k) = 0$, otherwise H is undefined. Hence H is either constant or undefined. Since our definition of H is finite explicit, depending only on n, it follows that there is a p.r. function $r(n)$ such that H is defined by the rth set of equations where $r = r(n)$. (This, of course, requires trivial but tedious verification.) Now, $\chi_g(n) = 1 \leftrightarrow 0 \in$ range of $f_r \leftrightarrow f_r(r+1) = 0 \leftrightarrow \chi(r+1) = 1$. Thus we have $\chi_g(n) = \chi(r(n) + 1)$ and χ_g is recursive in χ.

For a long while the question of whether unsolvable problems existed which were not of maximal unsolvability remained open. This problem, known as "Post's problem," was solved simultaneously by Friedberg and Mučnik in 1956. The proof involves a more complicated diagonalization procedure than that used to define F. Further investigations have led to a rather thorough analysis of degrees of unsolvability. Nevertheless, it is still true that no "natural" problem (such as Hilbert's tenth), has led to non-maximal degrees of unsolvability.

Many natural classes of problems have been investigated and shown to be unsolvable. One of these is the word problem for groups. Here we are given a finite number of "words" set equal to 1 of the form $a_1^{n_1} \cdots a_k^{n_k} = 1$ where n_i are integers, a_i are to be thought of as belonging to some group. Given another word W involving the same symbols, the problem is to determine if W necessarily equals 1 as a consequence of the given equations. Phrased differently, if W_i are the given words we can think of them as being in the free group G generated by the symbols a_i. Our problem then is to determine if W can be obtained from the W_i by multiplication, taking inverses, and conjugation. This problem has been shown to be unsolvable even in the case where there are only two letters used in the words. Hilbert's tenth problem, to determine whether a given Diophantine equation has a solution, has still not been shown unsolvable. If one allows variable exponents, it has been shown to be unsolvable.

The reader may wonder why our interpretation of the notion of a "problem" is so narrow as to only mean finding the characteristic function of the range of a p.r. function. The statement $A \equiv \forall k\ (\sim f(k) = n)$ for a p.r. f is a statement which can be verified for each k, and hence A represents the simplest type of statement which cannot be immediately decided. Thus there might be some reasonable hope that a clever technique could mechanically decide A. Also many natural problems are indeed of this form.

Though we have already mentioned this point, perhaps we should emphasize that statements such as the "Riemann Hypothesis may prove to be undecidable" can only have meaning with regard to particular axiom systems and can never have an absolute character as our results on unsolvable problems. Later in the course we shall prove the undecidability of the continuum hypothesis with respect to Zermelo-Fraenkel set theory, and in this case one does have good reason to feel that it remains undecidable in any "natural" system of axioms.

We end this section by mentioning a theorem which also asserts that no general recursive function of a certain type exists.

THEOREM. Let $f_1(x,y)$ and $f_2(x,y)$ be two functions on the integers such that the axioms of Z_1 hold if we define $x + y$ as $f_1(x,y)$ and $x \cdot y$ as $f_2(x,y)$. Then either the resulting model for Z_1 is isomorphic to the integers under the usual operations + and $\cdot$, or either f_1 or f_2 is not general recursive.

Proof. Let M be the model defined by f_1 and f_2. Assume these are recursive. If M is not isomorphic to the usual model, let $a_n = f_1(a_{n-1}, b)$ where b is the number "1" in M. Thus a_n are the "standard" integers. Since M is not standard, some c must be greater than all the a_n, i.e., for each n there is an x and $c = f_1(x, a_n)$. We know that we can explicitly define a non-recursive recursively enumerable set S in Z_1. This definition, when carried out in M, yields a set $S' \subseteq M$. Let p_n be the n-th prime in the model M. By the Chinese Remainder Theorem, which must be true in M, there is a y in M such that $y \equiv 0$ or 1 (mod p_n) for all $n < c$, according as $n \in S'$ or $\sim n \in S'$. For any "standard" integer n we shall now show how to tell recursively if $n \in S'$ or $\sim n \in S'$. Namely, we first locate p_n in M. Next, we successively test all the members of M to see if $p_n \cdot z$ equals y or y-1. In the first case $n \in S'$, in the second $\sim n \in S'$. Since f_1, f_2 are recursive this is a recursive procedure. However, we do not yet have a contradiction, since if N denotes the "standard" integers, we can not say that $N \cap S' = S$. Observe that if S is defined as the range of a particular p.r. function f, we always have $N \cap S' \supseteq S$, since if $n \in S$, $\exists\, m \in N$ such that if $f(m) = n$ and this equation must hold in M as well. We shall show in a lemma below that two recursively enumerable sets S and T can be defined, so that one can prove in Z_1 that $S \cap T = \emptyset$ and if U is a recursive set we do not have both $S \subseteq U$ and $U \cap T = \emptyset$. Since $S \cap T = \emptyset$ is provable in Z_1, $S' \cap T' = \emptyset$. But we have just shown that $S' \cap N = U$ is recursive and since $T' \supseteq T$, we have $S \subseteq U$, $U \cap T = \emptyset$ which completes the proof.

LEMMA. There are two explicitly defined recursively enumerable sets S and T, $S \cap T = \emptyset$, such that if U is recursive we do not have both $S \subseteq U$ and $U \cap T = \emptyset$.

Proof. Let φ_n enumerate all Turing machines. We define p.r. functions f and g as follows. Run through all possible computations of $\varphi_n(n+2)$ for all n. If no calculation is reached at the m-th stage or if $\varphi_n(n+2)$ is neither 0 or 1, put $f(m) = 0$, $g(m) = 1$. If $\varphi_n(n+2) = 0$ put $f(m) = n+2$, $g(m) = 1$. If $\varphi_n(n+2) = 1$, put $f(m) = 0$, $g(m) = n+2$. Let S be the range of f, T the range of g so clearly $S \cap T = \emptyset$. If φ_n is actually a function and $U_n = \{x \mid \varphi_n(x) = 1\}$ then either $n+2 \in S$ and $\sim n+2 \in U_n$, or $n+2 \in U_n$ and $n+2 \in T$ so that U_n cannot violate the lemma. Since U_n runs through all recursive sets, the lemma is proved.

On the other hand, the completeness theorem yields the existence of non-standard models. For example, we can adjoin the statements $c > 1$, $c > 2, \ldots$ to the axioms to get a consistent system. The proof of the completeness theorem then yields a model whose elements are the integers and $+$ and $\cdot$ are given by arithmetically definable functions since the proof uses only arithmetic ideas. The above theorem however says these functions cannot be recursive.

By a similar argument using Tarski's theorem on the undefinability of truth, and without using the above lemma one easily shows the following.

THEOREM. There is no non-standard model M for the integers in which $+$ and $\cdot$ are given by functions definable in Z_1 and such that all the true statements of Z_1 hold in M.

CHAPTER II

ZERMELO-FRAENKEL SET THEORY

1. AXIOMS

In Chapter I we discussed formal systems for elementary number theory. Certainly number theory is only a small part of mathematics as we know it and the question arises how to axiomatize all of conventional mathematics. The chief idea which one uses in passing to higher systems of mathematics is that of considering as a single object in the discussion, a set of objects previously considered. Sometimes the set concept occurs in the form of sequences, or functions, but these are only inessential variations. Thus, a real number is a sequence of rationals, a real-valued function is a set of ordered pairs of real numbers, an L^p space is a set of functions, etc. By analyzing mathematical arguments, logicians became convinced that the notion of "set" is the most fundamental concept of mathematics. This is not meant to detract from the fundamental character of the integers. Indeed a very reasonable position would be to accept the integers as primitive entities and then use sets to form higher entities. However, it can be shown that even the notion of an integer can be derived from the abstract notion of a set, and this is the approach we shall take. Thus in our system, all objects are sets. We do not postulate the existence of any more primitive objects. To guide his intuition, the reader should think of our universe as all sets which can be built up by successive collecting processes, starting from the empty set. Thus our system will be similar to the system Z_2 of Chapter I, except that we allow the formation of infinite sets. The integers will be defined much as in Z_2.

The first set of axioms for a general set theory was given by E. Zermelo in 1908. It was later developed by A. Fraenkel and is usually referred to as Zermelo-Fraenkel (ZF) set theory. This is perhaps the most natural version of set theory and is the one with which we shall be most concerned. As in Z_1 and Z_2, one of our axioms will be an axiom scheme which consists of infinitely many axioms. Another system of axioms which has only finitely many axioms, but is less natural, was developed by von Neumann, Bernays, and Gödel and is usually referred to as Gödel-Bernays (GB) set theory. This system will be discussed later. We now state the axioms for ZF set theory with some comments.

1. Axiom of Extensionality

$\forall x, y\ (\forall z(z \in x \leftrightarrow z \in y) \rightarrow x = y)$.

This says, as in Z_2, that a set is determined by its members.

Definition. $x \subseteq y \leftrightarrow \forall z(z \in x \rightarrow z \in y)$. $x \subset y \leftrightarrow x \subseteq y\ \&\ \sim x = y$. In words, $x \subseteq y$ means x is contained in y.

2. Axiom of the Null Set

$\exists x\ \forall y\ (\sim y \in x)$.

The set defined by this axiom is the empty or null set and we denote it by $\emptyset$.

3. Axiom of Unordered Pairs

$\forall x, y\ \exists z\ \forall w(w \in z \leftrightarrow w = x \vee w = y)$.

We denote the set z by $\{x,y\}$. Also $\{x\}$ is $\{x,x\}$ and we put $\langle x,y\rangle = \{\{x\}, \{x,y\}\}$. The set $\langle x,y\rangle$ is called the "ordered pair" of x and y and it is easily proved that $\langle x,y\rangle = \langle u,v\rangle$ implies $x = u$ and $y = v$. There is no special significance to the form of the definition of $\langle x,y\rangle$. It is just a convenient way of reducing the idea of ordered pairs to unordered pairs.

Definition. A function is a set f of ordered pairs such that $\langle x,y\rangle$ and $\langle x,z\rangle$ in $f \rightarrow y = z$. The set of x such that $\langle x,y\rangle \in f$ is called the domain. The set of y is the range. We say f maps into a set u, if the range of f is contained in u.

4. Axiom of the Sum Set or Union

$\forall x \; \exists y \; \forall z (z \in y \leftrightarrow \exists t (z \in t \;\&\; t \in x))$.

This says that y is the union of all the sets in x. In particular using Axiom 3 we can thus deduce that given x and y, there exists z such that $z = x \cup y$, i.e., $t \in z \leftrightarrow t \in x \vee t \in y$.

To motivate the next axiom, we remark that as in Z_2, if x is an integer the successor of x will be defined as $x \cup \{x\}$. Then the following axiom guarantees the existence of a set which contains all the integers and is thus infinite.

5. Axiom of Infinity

$\exists x \; (\emptyset \in x \;\&\; \forall y (y \in x \rightarrow y \cup \{y\} \in x))$.

In ZF, after we have defined integers, the principle of induction is just a single sentence, namely, that every set of integers has a least element. However, this still does not mean that the need for an infinite axiom scheme, as in Z_2, can be avoided. For, we have merely shifted the burden to an axiom which will guarantee the existence of sets to which the induction principle can be applied. This axiom is perhaps the most characteristic axiom of ZF. In its most naive form it says that every property defines a set. In such generality, we are led to a contradiction (e.g., the set of all x such that $\sim x \in x$). In mathematics we do define sets by properties but our intuition tells us to avoid certain properties which might lead us to "absurd" sets. One can restrict the type of properties allowed and obtain axioms of varying strength. We shall state the axiom in an extremely powerful form which goes beyond all applications of it in conventional mathematics. Nevertheless, it is carefully worded so as to avoid (presumably!) any contradiction. To state it, we first enumerate the countably many formulas in our system with at least two free variables, $A_n(x,y;t_1,\ldots,t_k)$ where k depends on n.

6_n. Axiom of Replacement

$\forall t_1,\ldots,t_k \; (\forall x \; \exists! \, y \; A_n(x,y;t_1,\ldots,t_k) \rightarrow \forall u \; \exists v \; B(u,v))$ where $B(u,v) \equiv \forall r \; (r \in v \leftrightarrow \exists s \; (s \in u \;\&\; A_n(s,r;t_1,\ldots,t_k)))$.

In other words, the axiom says that if for fixed $t_1,\ldots,t_k$, $A_n(x,y;t_1)$ defines y uniquely as a function of x, say $y = \varphi(x)$, then for each u the range of φ on u is a set. The intuition that says that 6_n does not lead to a contradiction is probably that v is prevented from being an absurd set by the fact that its cardinality, as we shall later define it, does not exceed that of u. The property A_n which defines φ, may however be extremely non-constructive in character, so that to verify $y = \varphi(x)$ may require answering a question about all sets. Thus axiom 6_n allows for the most non-constructive definitions. We shall later discuss a less powerful substitute for 6_n, which still suffices for most purposes. It should also be remarked that the role of the t_i is merely that of parameters in the definition of the function φ.

7. Axiom of the Power Set

$$\forall x\ \exists y\ \forall z\ (z \in y \leftrightarrow z \subseteq x).$$

This axiom says that there exists for each x the set y of all subsets of x. Although y is thus defined by a property, it is not covered by the Replacement Axiom because it is not given as the range of any function. Indeed, the cardinality of y will be greater than that of x so that this axiom allows us to construct higher cardinals. We shall later give a rigorous proof that axiom 7 cannot be derived from the other axioms.

8. Axiom of Choice

If $\alpha \to A_\alpha \neq \emptyset$ is a function defined for all $\alpha \in x$, then there exists another function $f(\alpha)$ for $\alpha \in x$, and $f(\alpha) \in A_\alpha$.

This is the well-known Axiom of Choice, which allows us to do an infinite amount of "choosing" even though we have no property which would define the choice function and thus enable us to use 6_n instead. As is well-known, this is an extremely useful axiom in modern mathematics, although its importance was not immediately understood.

9. Axiom of Regularity

$$\forall x\ \exists y\ (x = \emptyset \vee (y \in x\ \&\ \forall z\ (z \in x \to \sim z \in y))).$$

This axiom is a somewhat artifical one and we include it for technical reasons only. It is never used in conventional mathematics. It says that each non-empty set x, contains an element which is minimal with respect to the relation $\in$ (not $\subseteq$). Intuitively, we wish all our sets to be "built up" from $\emptyset$ and so we do not want to have infinite descending chains with respect to $\in$; rather all descending chains should end with $\emptyset$. In conventional mathematics, we only deal with integers, real numbers, functions, etc. and these are explicitly built up from $\emptyset$. This axiom explicitly prohibits $x \in x$, for example, since otherwise $\{x\}$ would not have a minimal element. The condition on $\in$ given by Axiom 9 is similar to the notion of a well-ordering, but of course $\in$ does not order all sets since we can have $x \neq y$, $\sim x \in y$ and $\sim y \in x$. One sometimes says that $\in$ is a "well-founding" relation or that all sets are "well-founded".

This completes our axioms. To stay totally within the formal system of ZF would mean using only the logical symbols together with $\in$. This of course is highly impractical and we have already used abbreviations. It should be emphasized that from this point on, the development of set theory is assumed to take place wholly within ZF, unless we are discussing the system ZF itself where we shall sometimes use still more powerful principles. In these cases we shall usually explicitly mention these higher principles. The reader should be reasonably convinced even on a first reading that the axioms easily encompass all of traditional mathematics.

2. DISCUSSION OF THE AXIOMS

We now give an informal discussion of the axioms although our results are precise and could be formalized. If the Axiom of Extensionality is dropped, the resulting system may contain atoms, i.e., sets x such that $\forall y\ (\sim y \in x)$ yet the sets x are different. Indeed, one possible view is that the integers are atoms and should not be viewed as sets. Even in this case, one might still wish to prevent the existence of unrestricted atoms. In any case, for the "genuine" sets, Extensionality holds and the other sets are merely harmless curiosities. Fraenkel and Mostowski have used atoms to obtain results about the Axiom of Choice, and we shall discuss these results in a later chapter.

The first really interesting axiom is the Axiom of Infinity. If we drop it, then we can take as a model for ZF the set M of all finite sets which can be built up from $\emptyset$. We have already discussed M in connection with Z_2 in I § 7. It is clear that M will be a model for the other axioms, since none of these lead out of the class of finite sets.

Hence we have shown that axiom 5 cannot be deduced from the other axioms since 5 is certainly false in the model of finite sets.

Axiom 6_n is an extremely powerful axiom since it allows forming a new set by putting together sets chosen from the entire universe of sets. A more restricted axiom which may be also more philosophically palatable is the following "Aussonderung" or "Separation" axiom. Here $A_n(z; t_1, \ldots, t_k)$ ranges over all formulas with at least one free variable.

$6'_n$. <u>Axiom of Separation</u>

$$\forall t_1, \ldots, t_k \ \forall x \ \exists y \ \forall z \ (z \in y \leftrightarrow z \in x \ \& \ A_n(z; t_1, \ldots, t_k)).$$

That is, for every x there is a set y of all the z in x satisfying the property A. To verify A for a particular z may involve asking a question about all sets, but the point is that the new set y is a subset of x. The axiom scheme $6'_n$ is easily derived from 6_n, by defining the function $\varphi(t) = t$ if t satisfies A and $\varphi(t) = x$, say, if t does not. The range of φ on x is then precisely the set $y' = y \cup \{x\}$. Thus $y' \cap x$ gives y. It is intuitively clear that 6_n is more powerful than $6'_n$ since 6_n allows us to build up as well as cut down. A rigorous proof is the following: For all x, let $P(x)$ denote the power set of x. Put $u_0 = \emptyset$, $u_n = P(u_{n-1})$ and $M = \bigcup_n u_n$, the union of the u_n so that M is the model of all finite sets which was previously discussed. Now put $C_0 = M$, $C_n = P(C_{n-1})$ and $N = \bigcup_n C_n$. It is clear then that all the axioms, with $6'_n$ replacing 6_n, hold in N, since if $x \in N$, so does $P(x) \in N$. However, the function $n \to C_n$ can be defined in ZF (we have just done it), and so if 6_n held in N, one would have that $\{C_0, C_1, \ldots\}$ is a member of N which is clearly absurd. Thus 6_n is false in N and hence $6'_n$ and the remaining axioms do not imply 6_n.

In the informal arguments we shall give, axiom 6_n will often be implicitly used. Whenever we define a set by saying "S is the set of all x such that...," we are of course using 6_n. Sometimes we implicitly use axiom 4 as well. For example, if we have sets A_n then to define the union of the A_n, we must first use 6_n to obtain $\{A_1, A_2, \ldots\}$ and then apply axiom 4 to that set to obtain $\bigcup_n A_n$. Also, it will always be true that the property used to define a set will be expressible in ZF. The reader will only need to verify that the terms used in the property have previously been defined within ZF.

We next consider the Power Set Axiom. For a set x, let us denote by Sx the sum-set of axiom 4. Also, we define $S_0x = x$, $S_{n+1}x = S(S_nx)$. To give a model in which the Power Set Axiom is false, the natural idea would be to consider the class of all countable sets since it is only through the power set that one can prove the existence of uncountable sets. However, a set x can be countable and yet Sx be uncountable. Let M be the set of all x such that S_nx is countable for all n. (We shall later show that from the axioms of ZF it follows that such a set M exists.) It is now easy to see that the axioms, with the exception of the Power Set Axiom, hold in M. The only difficulty is axiom 6_n. Now, it is clear that if $x_n \in M$ then $\{x_1, x_2, \ldots\}$ also belongs to M. Thus, φ is a function defined in M, then the range of φ on a set of M is a countable set all of whose members are in M and hence it is also in M. Thus 6_n holds in M. Clearly the Power Set Axiom fails in M, since in M all sets are countable, i.e., for x in M, $\exists y$ in M, such that y is a set of ordered pairs putting x into 1-1 correspondence with the integers (which will later be defined and are clearly in M). On the other hand, the Power Set Axiom implies the existence of uncountable sets. This example also shows the relevance of the Power Set Axiom for the proof of the existence of uncountable sets.

These elementary independence results are all contingent on the fact that ZF is consistent, since if ZF is inconsistent the axioms imply any statement. Of course the Incompleteness Theorem prevents a proof of Consis ZF being given in ZF. In the final analysis Consis ZF is essentially an article of faith and all the consistency and independence results we give are only <u>relative</u> consistency results depending on it.

3. ORDINAL NUMBERS

Ordinal numbers are important in set theory not only as generalizations of the natural numbers, but also play a key role in almost all investigations of the axiom system. This is due to the fact that the complexity of a given model is in part measured by the ordinals available to us in building up sets by means of a transfinite process. Intuitively, an ordinal stands for an equivalence class of well-ordered sets. As in all such definitions this leads to some awkwardness, since we must first restrict the domain of objects on which the equivalence relation is defined before we can make an identification. A much more satisfactory solution, proposed by von Neumann, is to pick a canonical representative from each equivalence class.

We have already defined function, domain and range. We give some more definitions for future use.

<u>Definition</u>. A relation R on a set X is a set of ordered pairs of elements of X.

$X \times Y$ is the set of all $\langle u,v\rangle$ such that $u \in X$ and $v \in Y$.

$\langle x_1,\dots,x_n\rangle = \langle x_1, \langle x_2,\dots,x_n\rangle\rangle$.

If f is a function, domain of f is x and $y \subseteq x$, then $f|y$ (f restricted to y) is the set of all $\langle u,v\rangle$ in f such that $u \in y$.

If $f(x) = f(y) \to x = y$, we define $f^{-1} = \{\langle x,y\rangle | \langle y,x\rangle \in f\}$.

The axioms easily justify the definitions. Thus, $X \times Y$ is a subset of $P(P(X \cup Y))$. We proceed to well-orderings.

<u>Definition</u>. A relation R on a set S, <u>orders</u> S if (writing $x < y$ in place of $\langle x,y\rangle$ in R),

i) $\forall x, y$ in S, one and only one of the following holds

$$x = y, \qquad x < y, \qquad y < x$$

ii) $x < y$ and $y < z \to x < z$.

We say R <u>well-orders</u> S if R orders S and if $B \subseteq S$, and B is not empty $\to \exists x(x \in B \;\&\; \forall y\ (y \in B \to \sim y < x))$.

Thus R well-orders S if every subset of S has a least element with respect to R. For $B \subseteq S$, we write $\sup B = x$, if x is the least element in S such that $y \in B \to y < x$. Also we write $x > y$ in place of $y < x$ and $x \leq y$ to mean $x = y \vee x < y$.

<u>Definition</u>. Let S be well-ordered by R. We say B is an initial segment of S if $x \in B$, $y < x \to y \in B$.

We observe that if B is an initial segment either B is all of S, or else $\exists x$ such that B is the set of all y such that $y < x$. For if $S - B$ is not empty let x be its least element. Then clearly if $y < x$, y is in B. On the other hand if $y \in B$ and $x < y$ then $x \in B$ since B is an initial segment, which contradicts $x \in S - B$. Hence $y < x$. Also, for any x, $\{y | y \leq x\}$ is an initial segment.

We say a function f from an ordered set S to an ordered set T is order-preserving if $x < y \rightarrow f(x) < f(y)$. Also, observe that a subset of a well-ordered set is well-ordered.

THEOREM. If S and T are well-ordered sets then either

i) $\exists$ a unique order-preserving function f from S onto an initial segment of T.

or ii) $\exists$ a unique order-preserving function from T onto an initial segment of S.

<u>Proof</u>. Let us call an order-preserving map from a well-ordered set onto an initial segment of another well-ordered set a "good" map. Now, if f is a good map we must have for all x, $f(x) = \sup\{f(y) | y < x\}$. For, since f is order-preserving $f(y) < f(x)$ if $y < x$ and hence $f(x) \geq t = \sup\{f(y) | y < x\}$. If $f(x) > t$, then t must be in the range of f, since the range is an initial segment, and $t = f(y)$ for some $y < x$ since f is order-preserving. This contradicts the definition of t and hence $f(x) = \sup\{f(y) | y < x\}$. We next observe for any two well-ordered sets S and T there can be at most one good map from S into T. For, let f and g be two such maps and let x be the least element in S such that $f(x) \neq g(x)$. Then $f(x) = \sup\{f(y) | y < x\} = \sup\{g(y) | y < x\} = g(x)$, a contradiction. The restriction of a good map f to an initial segment B is still a good map. For, it is order-preserving and if $x \in B$, $z < f(x)$, then $z = f(y)$ for some $y < x$ since f is good and $y \in B$ since B is an initial segment so the range of f on B is an initial segment. Let C now be defined as the set of all x in S such that there is a good map from the initial segment $\{y | y \leq x\}$ into T. If $x \in C$ and $y < x$ and f is a good map from $\{z | z \leq x\}$ into T, then f restricted to $\{z | z \leq y\}$ is also a good map so $y \in C$. Thus C is an initial segment. For $x \in C$, let f_x denote

the unique good map from $\{z | z \leq x\}$ into T. Clearly then $f_x(z) = f_y(z)$ if $z \leq x \leq y$. For all x in C, put $f(x) = f_x(x)$. If $x, y \in C$ and $x < y$ then $f(x) = f_x(x) = f_y(x) < f_y(y) = f(y)$ so f is order-preserving. If $x \in C$, $t < f(x)$, then $t < f_x(x)$ so $\exists y$, $y < x$ and $t = f_x(y) = f_y(y) = f(y)$. Thus f is a good map. If C is all of S we have proved the theorem. If $C \neq S$, let $t = \sup C$ and let C' be the range of f on C. If C' is all of T, then the inverse map of f is an order-preserving map from T onto an initial segment of S. If C' is not all of T, let $u = \sup C'$. If we put $f(t) = u$ we easily see that we have extended f to a map of $\{z | z \leq t\}$ onto an initial segment of T which is a contradiction and the theorem is proved.

THEOREM. If f is a good map from S into T and g is a good map from T into S, then both f and g are onto and they are inverse to one another.

<u>Proof</u>. Let f map S onto the initial segment B of T. Then f^{-1} is a good map from B to S. The restriction of g to B is also a good map from B to S, hence agrees with f^{-1}. But f^{-1} is onto all of S, so that since g is one-one we see that $B = T$ and $f^{-1} = g$.

<u>Definition</u>. If S and T are well-ordered sets we write $\tilde{S} \leq \tilde{T}$ if there is a good map from S into T. If this good map is not onto all of T we write $\tilde{S} < \tilde{T}$. If the good map is onto T we write $\tilde{S} = \tilde{T}$.

It is now clear that the notion of equality we have introduced is an equivalence relation and that $<$ is an order relation among the induced equivalence classes. (We shall actually not introduce any equivalence class and the fact that $<$ is an order relation merely means that given S, T one and only one of $\tilde{S} < \tilde{T}$, $\tilde{T} < \tilde{S}$, $\tilde{S} = \tilde{T}$ holds, etc.)

We now give some examples of well-ordered sets.

1. The set ω of integers under the usual ordering. (We have not yet defined the integers in ZF however.)
2. If S and T are disjoint well-ordered sets define $S + T$ as $S \cup T$ with the ordering $x < y \leftrightarrow x \in S$ and $y \in T$ or $x, y \in S$ and $x < y$ or $x, y \in T$ and $x < y$.
3. If S and T are well-ordered define a well-ordering on $S \times T$ by saying $\langle x,y \rangle < \langle t,u \rangle$ if $y < u$ or $y = u$ and $x < t$.

It is trivial to verify that these definitions preserve the equivalence classes of well-orderings defined by the $=$ relation. Observe that $S + T$ can be defined for S and T not disjoint by finding S' and T' disjoint such that $\tilde{S} = \tilde{S}'$ and $\tilde{T} = \tilde{T}'$. Ordinal arithmetic is a bit peculiar. Thus, $2 + \omega = \omega$, yet $\omega + 2 \neq \omega$. Also $2 \times \omega = \omega$, but $\omega \times 2 = \omega + \omega$. (Here 2 is the well-ordered set of 2 elements.) One can also define exponentiation but we shall wait until we have defined transfinite induction.

Definition. A set x is called transitive if $y \in x, z \in y \to z \in x$.

Definition. An ordinal is a set α which is well-ordered by $\in$ and which is transitive. We write $\mathrm{On}\ \alpha$ to mean α is an ordinal.

The point of this definition is that we are attempting to choose a canonical representative from each equivalence class of well-orderings. The most natural order relation to use is $\in$. One sees that the sets $\emptyset$, $\{\emptyset\}$, $\{\emptyset, \{\emptyset\}\}$, and in general if $n+1 = n \cup \{n\}$, the sets n, are ordinals. The second condition is necessary so that for example $\{\emptyset\}$ and $\{a\}$, for any set a, are not both taken as ordinals. With it, we shall prove that if two ordinals are equivalent as well-ordered sets, they are identical. We also observe that the Axiom of Regularity implies that any set ordered by $\in$ is well-ordered by it. However, we prefer to make our discussion independent of this axiom.

We observe some simple facts. An initial segment of an ordinal is an ordinal. For if I is an initial segment of α, I is well-ordered by $\in$ and if $x \in I$, $y \in x$, then y is less than x in the ordering of α (since $\in$ is the order relation), and hence $y \in I$, so I is transitive. Also if $\mathrm{On}\ \alpha$ and $x \in \alpha$, x is the set of all y in α such that y precedes x. This is true since $\in$ is the order relation and $x = \{y \mid y \in x\} = \{y \mid y \in x \ \&\ y \in \alpha\}$. If $\mathrm{On}\ \alpha$ and $x \in \alpha$ it follows that $\mathrm{On}\ x$. For, we know $x \subseteq \alpha$ so x is well-ordered by $\in$. Also, if $y \in x$ and $z \in y$ then $y \in \alpha$ and hence $z \in \alpha$ and since $\in$ is transitive on α, $z \in x$. Finally, if $\mathrm{On}\ \alpha$ then $\beta = \alpha \cup \{\alpha\}$ is an ordinal (the successor of α). For, if x, y are in β either $x, y \in \alpha$, or $x = y = \alpha$, or $x = \alpha$, $y \in \alpha$, or $y = \alpha$, $x \in \alpha$. In any case we see that either $x = y$, $x \in y$ or $y \in x$. Similarly we can verify the other conditions that β is an ordinal. Observe that $\beta - \alpha$ has only one member, α.

THEOREM. If S is a well-ordered set, there exist unique f and α such that On α and f is an order-preserving map of S onto α. That is, S is isomorphic to an ordinal.

<u>Proof</u>. We first note that $\{\emptyset\}$ is the unique ordinal with one element. For, if $x = \{a\}$ is an ordinal, $\sim a \in a$ since $\in$ orders x, and if $b \in a$, then $b \in x$ but $b \neq a$. Thus $a = \emptyset$. We have already shown that for any α, there is at most one isomorphism of S onto α. Let A be the set of all x in S such that $I_t = \{y | y \leq t\}$ for all $t \leq x$ has an isomorphism f_t onto a unique ordinal $\alpha(t)$. A is not empty since if x is the least element of S, I_x is isomorphic to $\{\emptyset\}$ and the latter is the unique such ordinal. Clearly A is an initial segment. Consider the function $f(x)$ defined for $x \in A$ by $f(x) = f_x(x)$. Using Replacement (axiom $6_n'$ does not suffice), we let α be the range of f. It is now easy to see that α is an ordinal and f is an isomorphism from A to α. Also if g is another isomorphism on an ordinal β, then for $x \in A$, g restricted to I_x is an isomorphism of I_x onto an initial segment J of β and On J, so that $g(x) = f(x)$ by the property defining A. Thus f and α are unique. Now if A is all of S we are done. If $A \neq S$, let $y = \sup A$ and define g on $A \cup \{g\}$ by putting $g(x) = f(x)$ for x in A and $g(y) = \alpha$. It is now clear g maps I_y isomorphically onto the ordinal $\alpha \cup \{\alpha\}$ which means $y \in A$ which is a contradiction.

COROLLARY. If On α and On β, then if $\tilde{\alpha} = \tilde{\beta}$, $\alpha = \beta$ and if $\tilde{\alpha} < \tilde{\beta}$ then $\alpha \in \beta$.

<u>Proof</u>. The first assertion is contained in the theorem. If $\tilde{\alpha} < \tilde{\beta}$, then there is an isomorphism of α onto a proper initial segement of β say $I = \{x < \gamma\}$ for some $\gamma \in \beta$. But I is an ordinal so $I = \alpha$ but we also know $I = \gamma$.

Thus, we have shown that <u>an ordinal is the set of all ordinals that precede it</u>.

THEOREM. If On α then $\beta = \alpha \cup \{\alpha\}$ is the least ordinal greater than α. We write $\beta = \alpha + 1$.

<u>Proof.</u> Exercise.

THEOREM. If S is a set of ordinals, there is a least α in S.

<u>Proof</u>. Let $\beta \in S$. The least element of $\beta+1 \cap S$ is the desired α.

THEOREM. If S is a set of ordinals, there is a least ordinal α such that $\beta \in S \to \beta < \alpha$. We write $\alpha = \sup S$.

<u>Proof</u>. Let γ be the sum-set of S. If $x, y, z \in \gamma$ then for α_1, α_2, α_3 in S, $x \in \alpha_1$, $y \in \alpha_2$, $z \in \alpha_3$ and the α_i are ordinals. If, for example, $\alpha_1 < \alpha_2 < \alpha_3$ then $x, y, z \in \alpha_3$ so that clearly $\in$ orders γ. If T is a subset of γ, say $\alpha \cap T \neq \emptyset$ and $\alpha \in S$. The least element of $\alpha \cap T$ is clearly the least element of T, so that γ is well-ordered and is also transitive. Thus γ is an ordinal and $\alpha \in S \to \alpha \in \gamma+1$. Further, it can be shown that γ is the least such ordinal, so that if $\beta = \gamma + 1$, β satisfies the theorem.

<u>Definition</u>. α is a successor if $\exists \beta\, (\alpha = \beta+1)$. α is a limit ordinal if $\alpha \neq 0$ and α is not a successor.

<u>Definition</u>. α is an integer if $\beta \leq \alpha \to \beta$ is a successor or 0.

THEOREM. There exists a limit ordinal.

<u>Proof</u>. This is essentially the content of the Axiom of Infinity, and could indeed be taken as a statement of it. Using our version of the axiom, let x be a set satisfying the Axiom of Infinity. Let y be the sup of all ordinals in x. Then since $\alpha \in x \to \alpha + 1 \in x$, y is a limit ordinal.

Now, if x is a limit ordinal, let ω be the least limit ordinal $\leq x$. Then if n is an integer, $n \in \omega$. Conversely if $y \in \omega$ then $z \leq y \to z$ is a successor or 0 so that y is an integer. Thus ω is the set of all integers.

THEOREM. (Mathematical Induction) If $x \subseteq \omega$ and $\emptyset \in x$ and $n \in x \rightarrow n + 1 \in x$ then $x = \omega$.

Proof. Exercise.

We now turn to transfinite induction, which is an extension of the usual notion of induction for integers. The object here is to define a function $f(\alpha)$ for ordinals α from a rule which tells how to define $f(\alpha)$ if $f(\beta)$ is known for all $\beta < \alpha$. The fact that we deal with well-ordered sets means that this can always be done uniquely. To state it formally, let $A_n(x,y;t_1,\ldots,t_k)$ run through all formulas in ZF with at least two free variables.

THEOREM. (Transfinite Induction) Let t_i be given and assume $\forall x\ \exists!\ y\ A_n(x,y;t_1,\ldots,t_k)$ so that A defines a function $y = \varphi(x)$. For every ordinal α and set z there is a unique function f defined on $\{\beta | \beta \leq \alpha\}$ (this set is actually $\alpha + 1$), such that $f(0) = z$ and for all $\beta \leq \alpha$, $f(\beta) = \varphi(h)$, where h is f restricted to β.

Proof. Let α and z be given. Let S be the set of all ordinals $\gamma \leq \alpha$ such that the theorem holds with γ in place of α, and let f_γ be the corresponding function. It is clear that if $\gamma_1 \in S$, $\gamma_2 < \gamma_1$ then $\gamma_2 \in S$ and f_{γ_2} is f_{γ_1} cut down to $\gamma_2 + 1$. Thus S is itself an ordinal. If $\alpha \in S$, we are done. If not, let γ_0 be sup S. Let g be defined on S by putting $g(\gamma) = f_\gamma(\gamma)$ for $\gamma < \gamma_0$. If we put $\delta = \varphi(g)$ and define $f(\gamma) = g(\gamma)$ for $\gamma < \gamma_0$ and $f(\gamma_0) = \delta$ it is clear that this f is the unique function satisfying our hypothesis for γ_0 which means $\gamma_0 \in S$, a contradiction.

In applications we do not necessarily explicit φ but it is always clear how φ can be formalized in ZF. We give some examples.

1. For fixed α, we define $\alpha + \beta$ by putting $\alpha + 0 = \alpha$, $\alpha + \beta = \sup\{\alpha + \gamma | \gamma < \beta\}$.

2. For fixed α, define $\alpha \cdot \beta$, by $\alpha \cdot 0 = 0$, $\alpha \cdot (\beta + 1) = \alpha \cdot \beta + \alpha$, $\alpha \cdot \beta = \sup\{\alpha \cdot \gamma | \gamma < \beta\}$ if β is a limit ordinal.

3. For fixed α, define α^β by $\alpha^0 = 1$, $\alpha^{\beta+1} = \alpha^\beta \cdot \alpha$, $\alpha^\beta = \sup\{\alpha^\gamma | \gamma < \beta\}$ if β is a limit ordinal.

An ordinal is called countable if it can be put in one-one correspondence with ω. If α is countable, then since $\beta < \alpha \leftrightarrow \beta \in \alpha$, we have an enumeration β_n of all ordinals less than α. It is easily proved that α, β countable implies $\alpha + \beta, \alpha \cdot \beta$ countable. Surprisingly, α, β countable implies α^β is countable. For if β is the least countable ordinal such that α^β is not countable, then if $\beta = \gamma+1$, $\alpha^\beta = \alpha^\gamma \cdot \alpha$ is countable, while if β is a limit ordinal and γ_n are all its predecessors, $\alpha^\beta = \sup\{\alpha^{\gamma_n}\}$ and hence is the union of countably many countable sets and hence is countable.

The countable ordinals can be extraordinarily complex. The ordinal ω^ω represents the well-ordered set of polynomials with integral coefficients ordered lexicographically, which is a fairly complicated ordering. The ordinal $\epsilon_0 = \text{Lim}\ \omega_n$ where $\omega_n = \omega^{\omega_{n-1}}$, $\omega_1 = \omega$ is already quite difficult to visualize directly. If we define ϵ_α in general by putting $\epsilon_{\alpha+1} = \text{Lim}\ \delta_n$ where $\delta_1 = \epsilon_\alpha + 1$ and $\delta_{n+1} = \omega^{\delta_n}$, and $\epsilon_\alpha = \sup\{\epsilon_\beta | \beta < \alpha\}$ if α is a limit ordinal, ordinals such as ϵ_{ϵ_0} are almost impossible to visualize directly. These examples are not too complicated in that the ordinals can be defined in fairly weak systems of set theory. Later we show how to associate a countable ordinal with any natural set of axioms for set theory, which will turn out to be beyond the power of that system to describe.

It is perhaps not generally known, but Cantor's stimulus to study set theory arose from countable ordinals. In studying the problem of the uniqueness of the expansion of a function as a trigonometric series, he considered the operation of taking a set E of real numbers and removing its discrete points to form the set E^*. The question then arose of how many times the $*$ operation need be applied to eventually reach the empty set (if ever). By considering sets of real numbers which are well-ordered under $<$ one can see that the $*$ operation may be repeated a transfinite number of times, depending on the well-ordering before the set is eventually emptied. It is curious that this approach of Cantor's was not successful in solving the original problem, namely to show that if a trigonometric series converges to zero except for a countable set, the series is identically zero.

4. CARDINAL NUMBERS

Definition. A and B have the same cardinality, or in symbols $\bar{\bar{A}} = \bar{\bar{B}}$ if there is a one-one function mapping A onto B.

This relation $\bar{\bar{A}} = \bar{\bar{B}}$ defines an equivalence relation.

Definition. $\bar{\bar{A}} \leq \bar{\bar{B}}$ if there is a one-one map of A into B.

THEOREM. (Cantor-Bernstein) $\bar{\bar{A}} \leq \bar{\bar{B}}$ and $\bar{\bar{B}} \leq \bar{\bar{A}} \to \bar{\bar{A}} = \bar{\bar{B}}$.

Proof. Let f map A one-one into B, and g map B one-one into A. We can clearly assume A and B are disjoint. Now, it is easy to see that $A \cup B$ is the disjoint union of sequences $S = \{\ldots, x_n, y_n, \ldots\}$ where the sequence may or may not terminate on the left, and $x_n \in A$, $y_n \in B$, $f(x_n) = y_n$, $g(y_n) = x_{n+1}$, and if x_n is in the range of g, then y_{n-1} appears and if y_n is the range of f, x_n appears. We can now define a one-one map φ from $S \cap A$ onto $S \cap B$. Namely, if S contains x_n for all negative n we put $\varphi(x_n) = y_n$. If a least n occurs and x_n occurs for that n, again put $\varphi(x_n) = y_n$. If a y_n is the last element on the left, put $\varphi(x_n) = y_{n-1}$. Then φ is a one-one map of A onto B. Observe that the Axiom of Choice is not used in this proof.

If $\bar{\bar{A}} \leq \bar{\bar{B}}$ and $\sim \bar{\bar{A}} = \bar{\bar{B}}$ we write $\bar{\bar{A}} < \bar{\bar{B}}$. Thus $\bar{\bar{A}} < \bar{\bar{B}}$ induces a partial ordering. If A and B are both given well-ordered sets we have already shown that there is a one-one map of either A into B or B into A. We shall now show using the Axiom of Choice, that every set can be well-ordered and hence $<$ induces an order relation among the equivalence classes defined by the relation $\bar{\bar{A}} = \bar{\bar{B}}$.

WELL ORDERING THEOREM. Every set S can be well-ordered.

Proof. From the Axiom of Choice applied to the identity function on P(S), it follows that there is a function f defined for all $x \subseteq S$, $x \neq \emptyset$ and such that $f(x) \in x$. Call a well-ordering of a set $T \subseteq S$, a good well ordering if for all $x \in T$, $x = f(S - \{y | y \in T \text{ \& } y < x\}$. What this means is that in the well-ordering, the "next" element is always obtained by applying the choice function to those that remain. It is quite natural that by successively choosing elements, we can eventually well-order all of S.

Fact. If two subsets T_1 and T_2 have good well-orderings $<_1$ and $<_2$, then one is included in the other and is indeed an initial segment of the other and its ordering is the restriction of the other ordering. In particular there is at most one good ordering on a set T.

For, assume φ is an order preserving map of T_1 onto an initial segment of T_2. (Otherwise, we reverse the roles of T_1 and T_2.) Let x be the least element in T_1 such that $\varphi(x) \neq x$. Since φ maps onto an initial segment $\varphi(x) = \sup\{\varphi(y) | y <_1 x\}$ where sup refers to $<_2$, and since $<_2$ is good, $\varphi(x) = f(S - \{\varphi(y) | y <_1 x\}) = f(S - \{y | y <_1 x\}) = x$, since $<_1$ is good. Thus $\varphi(x) = x$ for all x and we have proved our assertion.

Now, let T be the union of all T_α such that T_α has a good well-ordering. Define $x < y$ for x, y in T, if $x, y \in T_\alpha$ for some α and $x < y$ in the unique good ordering on T_α. If $x \in T_\alpha$, $y \in T_\beta$, then if T_α is an initial segment of T_β, $x, y \in T_\beta$ so the ordering is defined and even uniquely defined for all x, y in T. Since the T_α are an "expanding" set of initial segments, one trivially verifies $<$ is a well-ordering. Furthermore $<$ is a good well-ordering. For if $x \in T_\alpha$, $\{y | y < x\} = \{y | y \in T_\alpha$ and $y <_\alpha x\}$ so $x = f(S - \{y | y < x\})$ since $<_\alpha$ is good. If T is not all of S, we let $x_0 = f(S-T)$, put $T_0 = T \cup \{x_0\}$ and define an ordering in T_0 extending that of T and putting $x < x_0$ if $x \in T$. This is clearly a good well-ordering and cannot be one of the original T_α. Thus $T = S$ and S has a good well-ordering.

It is trivial to observe that the well-ordering principle implies the Axiom of Choice since if we well-order $\bigcup_\alpha A_\alpha$, defining $f(\alpha)$ as the least element in A_α gives a choice function.

Definition. A cardinal is an ordinal α such that $\beta < \alpha \rightarrow \bar{\bar{\beta}} < \bar{\bar{\alpha}}$.

A cardinal is thus the least ordinal of that cardinality. Thus since every set can be well-ordered and thus has the same cardinality as some ordinal, we have picked out for each equivalence class of sets of the same cardinality a unique representative. One of the first questions Cantor considered was whether there are arbitrarily large cardinals. The answer is contained in the following theorem.

CANTOR'S THEOREM. For any A, $\overline{\overline{A}} < \overline{\overline{P(A)}}$.

Proof. Clearly $\overline{\overline{A}} \leq \overline{\overline{P(A)}}$. Assume φ maps A onto $P(A)$. Let $z = \{x \in A \mid \sim x \in \varphi(x)\}$. Then $z \subseteq A$ and if $z = \varphi(y)$ for $y \in A$ then $y \in z \rightarrow \sim y \in \varphi(y) \rightarrow \sim y \in z$ and $\sim y \in x \rightarrow y \in \varphi(y) \rightarrow y \in z$, which is clearly impossible.

It now follows that we can define by transfinite induction a function $\alpha \rightarrow \aleph_\alpha$ from ordinals to cardinals such that $\aleph_\alpha$ is the least infinite cardinal greater than that $\aleph_\beta$ for all $\beta < \alpha$. By Cantor's theorem we know that one can always find a cardinal greater than any given ordinal so that the definition is valid. It is clear, by induction, that $\aleph_\alpha \geq \alpha$. Thus it follows that $\aleph_\alpha$ must run through arbitrarily large cardinals and hence through all cardinals. This proof that $\aleph_\alpha$ exists uses the Axiom of Choice since we need to know that every set has the same cardinality as some ordinal. However, one can also prove without the Axiom of Choice that given any ordinal, there is another ordinal of greater cardinality, and thus that we can define $\aleph_\alpha$. Namely, if α is an ordinal consider all the well-orderings on α. They form a set since each well-ordering belongs to $P^3(\alpha)$. Consider the set of all equivalence classes of orderings under isomorphism. This set (a subset of $P^4(\alpha)$), is itself well-ordered in the natural way and it is easy to see that the ordinal β with which it is isomorphic is the first cardinal greater than α.

Let C denote the set $P(\omega)$. We know C is an uncountable set. Cantor considered the question of determining the position of C in the transfinite series of the $\aleph_\alpha$. Since $\overline{\overline{C}} > \aleph_0$ the simplest conjecture is

CANTOR'S CONTINUUM HYPOTHESIS. $\overline{\overline{C}} = \aleph_1$.

More generally one can conjecture the

GENERALIZED CONTINUUM HYPOTHESIS. $\overline{\overline{P(\aleph_\alpha)}} = \aleph_{\alpha+1}$

One often writes 2^A for the cardinality of $P(A)$ in analogy with finite sets.

The cardinals have their own arithmetic. Namely, if A and B are disjoint we put $\overline{\overline{A}} + \overline{\overline{B}} = \overline{\overline{A \cup B}}$ and $\overline{\overline{A}} \cdot \overline{\overline{B}} = \overline{\overline{A \times B}}$. However, this arithmetic is rather trivial since one can prove that if A and B are infinite $\overline{\overline{A}} + \overline{\overline{B}} = \overline{\overline{A}} \cdot \overline{\overline{B}} = \max(\overline{\overline{A}}, \overline{\overline{B}})$. The proof uses the Axiom of Choice and we omit it.

5. THE AXIOM OF REGULARITY

In this section we discuss the role of the Axiom of Regularity. In particular, we shall show that adding it to the other axioms does not lead to a contradiction if the latter are themselves consistent. This proof is especially interesting since it will serve as an introduction to the ideas behind Gödel's consistency proof for the Axiom of Choice and the Generalized Continuum Hypothesis.

We recall that Regularity simplifies our definition of an ordinal number since we can now omit the condition that ϵ well-orders α and merely say ϵ orders α. Recall also that a set x is transitive if $y \in x$ and $z \in y \rightarrow z \in x$.

Definition. A rank function on a transitive set A is an ordinal-valued function r on A such that for all $x \in A$, $r(x) = \sup\{r(y) \mid y \in x\}$.

LEMMA. Let A_1, A_2 be transitive, r_1 and r_2 rank functions on A_1 and A_2 respectively. Then $A_1 \cap A_2$ is transitive and $r_1 = r_2$ on $A_1 \cap A_2$.

Proof. First, $x \in A_1 \cap A_2$, $y \in x \rightarrow y \in A_1$ since A_1 is transitive and $y \in A_2$ since A_2 is transitive. So $y \in A_1 \cap A_2$ and $A_1 \cap A_2$ is transitive. Let α be the least ordinal such that $\exists x$ in $A_1 \cap A_2$ and $r_1(x) = \alpha$, $r_1(x) \neq r_2(x)$.

If $y \in x$, $r_1(y) < \alpha$ so $r_1(y) = r_2(y)$. Now $r_2(x) = \sup\{r_2(y) \mid y \epsilon x\}$ $= \sup\{r_1(y) \mid y < x\} = r_1(x)$, a contradiction.

The rank of x tells us how many times we must repeat the collecting process starting from $\emptyset$ to arrive at x.

Definition. A set x is well-founded if x belongs to some transitive set A for which a rank function exists.

By the lemma, the rank of x in any transitive set A such that $x \in A$ is unique and we can thus write rank x, with no ambiguity.

Definition. Let $S_0 x = \{x\}$, $S_{n+1} x$ = the sum-set of $S_n x$, $S_\omega x = \bigcup_{n=0}^{\infty} S_n x$.

THEOREM. $x \in S_\omega x$ and $S_\omega x$ is transitive. If $x \in A$ and A is transitive, $S_\omega x \subseteq A$.

<u>Proof</u>. If $y \in S_\omega x$, then $y \in S_n x$ for some n. Thus if $z \in y$, then $z \in S_{n+1} x$, so $z \in S_\omega x$. Also if A is transitive, $x \in A$, then $S_0 x \subseteq A$. If $S_n x \subseteq A$, then the transitivity of A means $S_{n+1} x \subseteq A$, so $S_\omega x \subseteq A$.

Using Regularity we now prove

THEOREM. Every set x is well-founded.

<u>Proof</u>. Assume x is not well-founded. In $S_\omega x$, by Regularity, there exists a set y such that y is not well-founded, and y is the "minimal" element with respect to $\in$ in $S_\omega x$ having this property. Hence $t \in y \to t$ is well-founded. Let $A = S_\omega y$. Then $A = \{y\} \cup \bigcup_{t \in y} S_\omega t$. For each $t \in y$, there is a rank function on $S_\omega t$. By the Lemma, if we put $r(z) = \text{rank } z$ we have a rank function on $\bigcup_{t \in y} S_\omega t$. Now if we set $r(y) = \sup\{r(z)\}$ we clearly have a rank function on A, so that y is well-founded.

COROLLARY. Every transitive set A has a rank function.

<u>Proof</u>. The function rank x is such a rank function.

We note for future reference that for each ordinal α, we can prove the existence of the set of sets of rank α. Namely, by transfinite induction define $S(\alpha) = P(\bigcup_{\beta < \alpha} S(\beta))$.

<u>From now on in this section we do not assume the Axiom of Regularity unless otherwise mentioned</u>.

LEMMA. If y is well-founded for every $y \in x$, then x is well-founded.

<u>Proof</u>. Consider $S_\omega x$. As before we can define a rank function on $\bigcup_{y \in x} S_\omega y$ by putting $r(t) = \text{rank } t$, since $S_\omega y$ being the smallest transitive set containing y, must have a rank function. If we put $r(x) = \sup\{r(y) \mid y \in x\}$, we obtain a rank function on the set $S_\omega x$ itself.

LEMMA. All ordinals α are well-founded and rank $\alpha = \alpha$.

Proof. Clear by the previous lemma and transfinite induction.

LEMMA. If x is well-founded, so is P(X).

Proof. Exercise.

Our next goal is to show that the Axiom of Regularity is consistent with the other axioms, if these are consistent themselves. The proof is very simple in principle but there are some subtle points which merit attention. By ZF we shall now mean ZF without Regularity and either with or without the Axiom of Choice since this does not affect the discussion. We begin with some general remarks about consistency proofs. If we are to work in ZF we cannot prove the consistency of ZF, hence certainly not the consistency of ZF + Regularity. Therefore we shall prove a relative consistency statement, namely Consis ZF $\rightarrow$ Consis ZF + Regularity. Since this is a statement in number theory, we can hope to prove it in Z_1 (or Z_2) and indeed the proof we give can be formalized entirely in Z_1, although we shall at first give the proof informally.

Let M be a model for some axiom system A in a given formal language. One may be interested in the subset N of M consisting of all x in M satisfying some property P(x) of the language. As will happen in our case, we may be able to show that N has certain properties, independently of the model M, but only as a consequence of the fact that M is a model for the axioms A. In such a case one can avoid speaking about models entirely and formulate a result in the formal system. For the sake of simplicity, we assume that our formal language has no constant symbols.

Definition. If F is a formula, then F_P denotes the formula obtained from F by adjoining to every variable x the condition P(x). That is, in building F_P, each $\exists x\, B$ becomes $\exists x[P(x) \,\&\, B]$ and each $\forall x\, B$ becomes $\forall x[P(x) \rightarrow B]$. We say F_P is F relativized to the condition P(x).

Now, we need the following obvious but tedious lemma. We assume that $\exists x\, P(x)$.

LEMMA. If F is a valid formula in the formal language, so is F_P.

A rigorous proof would require considering every rule of the Predicate Calculus and we would then have a purely combinatorial proof of the lemma. Intuitively the lemma is obvious, since if F is valid it holds for every set S in which relations and constants corresponding to the formal language are defined and hence for the subset $\{x \mid x \in S \ \& \ P(x)\}$. Thus F_P is true in every model and hence is valid.

Now we return to regularity. Let $P(x)$ be the condition "x is well-founded".

THEOREM. If A is an axiom of ZF, then one can prove in ZF the statement A_P.

In simpler terms we can say that <u>the well-founded sets are a model for</u> ZF.

Our proof of the theorem will be entirely combinatorial in that (in principle), we have a simple algorithm for producing a proof of A_P. Now, since $\emptyset$ is well-founded the Axiom of the Null Set holds. Extensionality clearly holds for the well-founded sets since if $x \in y$ and y is well-founded so is x. If x and y are well-founded so is $\{x,y\}$. If x is well-founded so are the sum-set of x and the power set of x. The set ω is well-founded, so the Axiom of Infinity holds. If $f(\alpha)$ is a choice function for the sets A_α and if the function $\alpha \to A_\alpha$ is well-founded so is the function $\alpha \to f(\alpha)$. The Replacement Axiom is proved as follows. Let $y = \varphi(x)$ be a function which is defined for well-founded sets and taking as values only well-founded set. Then the range of φ on any set consists only of well-founded sets and hence is itself well-founded. Thus we have proved the theorem.

THEOREM. The Axiom of Regularity, relativized to the well-founded sets, is provable in ZF.

<u>Proof</u>. Exercise.

THEOREM. If ZF is consistent, so is ZF + Axiom of Regularity.

Proof. Let R denote the regularity axiom. If ZF + R is inconsistent there is a valid statement of the form $R \,\&\, A_1 \,\&\, \cdots \,\&\, A_n \rightarrow B \,\&\, \sim B$ where A_i are axioms of ZF. But then $R_P \,\&\, A_{1,P} \,\&\, \cdots \,\&\, A_{n,P} \rightarrow B_P \,\&\sim B_P$ is also valid. Now, R_P and $A_{i,P}$ are all provable in ZF so that we have that the axioms of ZF $\rightarrow B_P \,\&\, \sim B_P$ which means ZF is inconsistent.

A sub-model which is characterized by a certain property is known as an "inner-model". Notice that in our inner model of well-founded sets we did not change the ϵ-relation. The method of inner models is used in the classical consistency proofs for non-Euclidean Geometry which imbed models for non-Euclidean geometry in Euclidean space. The method was also used by Gödel to give a proof of the consistency of the Continuum Hypothesis which we shall present in the next chapter.

The Axiom of Regularity shows $\sim x \in x$ and that there does not exist a sequence x_n such that $x_{n+1} \in x_n$ etc. However, the reason one never needs to quote this axiom in the conventional parts of mathematics is simply because the normal sets one deals with such as integers, reals, etc. are easily seen to be well-founded sets. If non-well-founded sets exist they must be rather artificial objects. Let us show how one can construct a "universe" with such sets. This argument will show that the Axiom of Regularity cannot be proved from the other axioms, so that in conjunction with our previous result we see that it is an independent axiom.

Let $x_1, x_2, \ldots$ be formal symbols and we formally define $x_{n+1} \in x_n$ and $\sim x_i \in x_j$ if $i \neq j+1$. Let $R(0) = \{x_1, x_2, \ldots\}$. Define $R(\alpha)$ by induction as the power set of $\bigcup_{\beta < \alpha} R(\beta)$. The objects in $R(\alpha)$ for $\alpha > 0$ are sets, whereas $R(0)$ consists only of the x_i. Now, among the members of the sets $R(\alpha)$ we define an ϵ-relation as follows: If u is a set, $v \in u$ if v is a member of u; if $u = x_i$, $v \in u$ if $v = x_{i+1}$. One can verify the axioms quite easily. Extensionality holds even for the x_i since if $i < j$, $x_{j+1} \in x_j$ but $\sim x_{j+1} \in x_i$. A rigorous proof requires defining the "formal symbols" x_i in ZF. Further examples of the above type can be given in which, for example the Axiom of Choice is false.

We now give a theorem, proved with regularity, which we shall have occasion to use several times.

THEOREM. Let A be a set which is extensional, i.e., $x, y \in A$, $x \neq y \rightarrow \exists z$ in A such that either $z \in x \,\&\, \sim z \in y$ or $z \in y \,\&\, \sim z \in x$. Then there exists a unique <u>transitive</u> set A' and a unique 1-1 map φ from A onto A' such that φ is an $\in$-isomorphism, i.e., $x \in y \longleftrightarrow \varphi(x) \in \varphi(y)$.

<u>Proof</u>. Define $\varphi(x)$ by induction on rank x. If rank $x = 0$, put $\varphi(x) = \emptyset$. If φ has been defined for all y such that rank $y < \alpha$, then if $x \in A$ and rank $x = \alpha$, put $\varphi(x) = \{\varphi(y) | y \in x \,\&\, y \in A\}$. Let $A' =$ range of φ. Clearly if $x \in y$ then $\varphi(x) \in \varphi(y)$. We first show that φ is 1-1, i.e., if $x \neq y$, $\varphi(x) \neq \varphi(y)$. We proceed by induction on $\alpha = \max(\text{rank } x, \text{rank } y)$. If $\alpha = 0$, then $x = y = \emptyset$. If the assertion is true for all $\beta < \alpha$, then if $x \neq y$, $\exists z \in A$ and say $z \in x$ and $\sim z \in y$. Then $\varphi(z) \in \varphi(x)$. If $\varphi(z) \in \varphi(y)$ then for some $w \in y$, $\varphi(z) = \varphi(w)$. But $\max(\text{rank } z, \text{rank } w) < \alpha$, so $z = w$, a contradiction. Now, if $\varphi(x) \in \varphi(y)$, then $\exists w$ in $A(w \in y \,\&\, \varphi(x) = \varphi(w))$. Hence $x = w$, and $x \in y$. Thus φ is an $\in$-isomorphism. Finally since $x \in \varphi(y) \rightarrow \exists w$ in A and $w \in y$, $x = \varphi(w)$, this shows that A' is transitive.

If there were two such functions, φ_1 and φ_2, let α be the least ordinal such that $\exists x$, rank $x = \alpha$ and $\varphi_1(x) \neq \varphi_2(x)$. Assume $\exists y$, $y \in \varphi_1(x) \,\&\, \sim y \in \varphi_2(x)$. Since the range of φ is transitive, $\exists z \in x$ and $y = \varphi_1(z)$. Now rank $z <$ rank $x = \alpha$, so $\varphi_1(z) = \varphi_2(z)$ and so $\varphi_2(z) \in \varphi_2(x)$ or $y \in \varphi_2(x)$, a contradiction.

On p. 56, in our discussion of the independence of the Power Set axiom, we assumed there was a set of all x such that $S_n x$ for all n is countable. The reader can easily verify that this is merely the set of all sets of countable rank and thus our argument was justified.

6. <u>THE SYSTEM OF GÖDEL-BERNAYS</u>

The axioms of ZF are infinite in number, because of the Axiom of Replacement. We shall later show that no finite number of axioms implies the full system of axioms. Although there are an infinite number of statements in the Axiom of Replacement, these statements are generated by a simple rule which can be described in a finite number of terms. Thus, it would seem plausible that one can axiomatize this construction and thus be reduced to a finite set of axioms. This is precisely what the axioms of Gödel-Bernays do.

The Axiom of Replacement says that if $y = \varphi(x)$ is a function, defined by some "property", the range of φ on any set u exists. The infinite number of statements are necessary in order to say what we mean by a "property". Thus we introduce a new primitive object called a "class". The intuitive idea which the reader should have in mind is that a class is the "collection" of all sets x satisfying a certain property $A(x)$. For example a trivial property such as $\sim x \in x$ determines the universal class. Of course, a set should certainly be considered a class since for a given set x the condition $y \in x$ is a property of y which defines x. The Axiom of Replacement can now be stated as a <u>single</u> axiom about classes which determine functions. What we also need, however, are axioms which allow us to show that any property gives rise to a class. Since every statement is constructed by finitely many applications of certain elementary processes such as conjunction, adding quantifiers etc., this can be described by finitely many axioms.

Two types of variables are thus used in GB. Upper case letters denote class variables and lower case letters, set variables. Equivalently we can regard our system as having only class variables and a unary relation $\mathfrak{M}(X)$, meaning "X is a set" (German: Menge), as well as the $\in$-relation. The axioms are as follows.

<u>Axioms for Sets</u>

1. $Y \in X \to Y$ is a set $(\mathfrak{M}(Y))$.
2. $X = Y \leftrightarrow \forall u\, (u \in X \leftrightarrow u \in Y)$.
3. $\exists x\ \forall y\, (\sim y \in x)$.
4. $\forall x\ \forall y\ \exists z\ \ \forall w (w \in z \leftrightarrow w = x \vee w = y)$.
5. $\forall x\ \exists y\ \forall z\, (z \in y \leftrightarrow \exists w (z \in w \,\&\, w \in x))$.
6. $\exists x\, (\emptyset \in x \,\&\, \forall y (y \in x \to y \cup \{y\} \in x))$.
7. $\forall x\ \exists y\ \forall z\, (z \in y \leftrightarrow z \subseteq x)$.

These axioms are the usual axioms for sets with the exception of 2, extensionality for classes and 1, which says that a class has only sets as member; this is consistent with our idea that a class is a "collection" of sets.

8. Axiom of Replacement

$\forall X((\forall u\ \exists!\ v\ \langle u,v\rangle \in X) \to \forall u\ \exists v$ (v is the range of the function defined by X on u)). The phrase written in words can be stated as $\forall t(t \in v \leftrightarrow \exists w(w \in u\ \&\ \langle w,t\rangle \in X))$.

<u>Axioms for Class Formations</u>

9. $\exists X\, \forall a(a \in X \leftrightarrow \exists b\, \exists c(a = \langle b,c\rangle\ \&\ b \in c))$

10. $\forall X\, \forall Y\, \exists Z\, \forall u(u \in Z \leftrightarrow u \in X\ \&\ u \in Y)$

11. $\forall X\, \exists Y\, \forall u(u \in Y \leftrightarrow \sim u \in X)$

12. $\forall X\, \exists Y\, \forall u(u \in Y \leftrightarrow \exists v(\langle v,u\rangle \in X))$

13. $\forall X\, \exists Y\, \forall u(u \in Y \leftrightarrow \exists r\, \exists s(u = \langle r,s\rangle\ \&\ s \in X)$

14. $\forall X\, \exists Y\, \forall a(a \in Y \leftrightarrow \exists b,c(\langle b,c\rangle = a\ \&\ \langle c,b\rangle \in X))$

15. $\forall X\, \exists Y\, \forall u(u \in Y \leftrightarrow \exists a,b,c(\langle a,b,c\rangle \in X\ \&\ \langle b,c,a\rangle \in Y\ \&\ \langle b,c,a\rangle = u))$

16. $\forall X\, \exists Y\, \forall u(u \in Y \leftrightarrow \exists a,b,c(\langle a,b,c\rangle \in X\ \&\ \langle a,c,b\rangle \in y\ \&\ \langle a,c,b\rangle = u))$

Observe that if we start with classes corresponding to properties, the classes which are asserted to exist by the above axioms also correspond to properties. We shall soon prove that for every property we can show the existence of a corresponding class.

17. Axiom of Choice

$$\exists X\ \forall a(a \neq \emptyset \to \exists!\ u(u \in a\ \&\ \langle a, u\rangle \in X)\} .$$

This is a rather strong version which says there is a class which picks out one member from each set.

18. Axiom of Regularity

$$\forall X\ (X \neq \emptyset \to \exists u(u \in X\ \&\ \forall y(y \in X \to \sim y \in u)) .$$

As we have said, the axioms of GB, with the exception of 17, are all valid in ZF if we replace the notion of "class" by that of "property". Axiom 17 does not immediately follow from ZF, because it is not clear how to give a rule in the language of ZF which picks out a unique member from every set. Indeed, one can show that such a strong form of the Axiom of Choice cannot be proved in ZF.

Let $A_n(t_1,\ldots,t_k;x_1,\ldots,x_\ell)$ run through all the formulas of ZF with free variables divided into two groups as indicated. The following then holds in GB.

THEOREM.

$$\forall t_1,\ldots,t_k\ \exists X\ \forall u(u \in X \leftrightarrow \exists x_1,\ldots,x_\ell(u = \langle x_1,\ldots,x_\ell\rangle \ \&\ A_n(t_i,x_j)))\ .$$

That is, any relation involving ℓ variables gives rise to a class of all ℓ-tuples satisfying that relation.

In the theorem we do not assume that A actually formally contains all the variables x_i, since we can always assume that A depends trivially on x_i. Now assume the theorem is true for all formulas of less than a given length, independently of the number of formal variables x_i which are used. We then have to consider the cases that A is of the form $B \,\&\, C$, $\sim B$ or $\exists u\, B(t_1,\ldots,t_k;u,x_1,\ldots,x_\ell)$. The first two cases are covered by axioms 10 and 11. To handle the third case, let Y be the class of all $\langle u,x_1,\ldots,x_\ell\rangle$ satisfying B for particular choices of the t_i. Then axiom 12 yields the existence of the required class X. It remains only to consider the case of the primitive formulas. Since the symbol $=$ can be avoided by replacing $x = y$ by $\forall z(z \in x \leftrightarrow z \in y)$, the only primitive formulas we have to consider are $x_i \in x_j$, $x_i \in t_j$, $t_i \in x_j$, and $t_i \in t_j$. Now from 10 and 11 we deduce the existence of a universal class X such that $\forall u(u \in X)$. From 13, we then deduce the existence of the class of all r-tuples of sets, for any r. The case $t_i \in t_j$ can now be disposed of since the class X is either the class of all ℓ-tuples or is the empty set. Consider the case $x_i \in x_j$ and assume first that $i < j$. By 9, the class of all $\langle x_i,x_j\rangle$ such that $x_i \in x_j$ exists. By 13, we can take the direct product of this class and the class of all $(\ell\text{-}j)$-tuples to deduce that the class of all $\langle\langle x_{j+1},\ldots,x_\ell\rangle, \langle x_i,x_j\rangle\rangle$ where $x_i \in x_j$, exists, and then by 14 the existence of the class of all $\langle\langle x_i,x_j\rangle, \langle x_{j+1},\ldots,x_\ell\rangle\rangle$ such that $x_i \in x_j$. Using 15 and recalling our inductive definition of r-tuples we obtain the existence of the class of all $\langle x_i,x_j,x_{j+1},\ldots,x_\ell\rangle$ and that $x_i \in x_j$. By repeating the same type of arguments, using 16 as well, we can finally show the existence of the class of all $\langle x_1,\ldots,x_\ell\rangle$ such that $x_i \in x_j$. The case $j < i$ follows easily as well as the case $i = j$. The class of all $\langle x_1,\ldots,x_\ell\rangle$ such that $x_i \in t_j$ is obtained by using axiom 13 to take the direct product of the set t_i with ℓ-1 copies of the universal class. We are left only with the case $t_i \in x_j$. The class $\{\langle t_i,x\rangle | \text{all } x\}$ exists since it is the direct product of $\{t_i\}$ and the universal class. Intersecting this with the class of axiom 9 yields the class of all $\{\langle t_j,x\rangle\ t_j \in x\}$. Axioms 12 and 14 now give the class of all $\{x | t_j \in x\}$. The direct product construction now gives the class of all $\{\langle x_1,\ldots,x_\ell\rangle | t_i \in x_j\}$ and the proof of the theorem is completed.

COROLLARY. $\forall t_i\ \exists X\ \forall u(u \in X \longleftrightarrow A_n(t_i,u))$.

That is, every property determines a class.

THEOREM. Every theorem of ZF is a theorem of GB.

Proof. We only need to show that the Replacement Axiom holds in GB. If $A_n(x,y;t_1,\ldots,t_n)$ is a formula in ZF which for particular t_i defines y as a function, $\varphi(x)$, of x, then the previous theorem gives the existence of a class X which is the graph of φ. Applying Axiom 8 then yields the Replacement Axiom.

To state the converse, let us assume that the Axiom of Choice in GB is replaced by the form of the axiom used in ZF. The theorem that follows can actually be proved without any such modification but the proof requires the method of "forcing" to be introduced in Chapter IV.

THEOREM. Any theorem in GB which speaks only about sets is a theorem in ZF.

Proof. Assume T is a theorem in GB but not in ZF. Then by the Completeness Theorem there is a countable model M for ZF in which T is false. Let $A_n(x;y_1,\ldots,y_k)$ enumerate all the formulas of ZF. Let $S_n(\bar{y}_1,\ldots,\bar{y}_k)$ for $\bar{y}_i \in M$ denote the set $\{x | x \in M \ \&\ A_{n,M}(x,\bar{y}_1,\ldots,\bar{y}_k)\}$. Let M' now be the union of M and the sets S_n with the convention that if for some $x \in M$, $\forall y(y \in M \rightarrow (y \in S_n \longleftrightarrow y \in x))$ then S_n and x are to be identified. It is now clear that M' is a model for GB with its elements considered as classes and those classes which are sets correspond precisely to the original members of M. Thus T is true in M' and hence true in M, since there are no new "sets" in M', which is a contradiction.

COROLLARY. ZF and GB are equiconsistent, i.e.,

$$\text{Consis(ZF)} \leftrightarrow \text{Consis(GB)}.$$

The proof of the above theorem and corollary could be done independently of any talk concerning models by a proof analysis. One could show that any proof in GB of a theorem involving only sets, requires the existence of only finitely many classes. The use of these particular classes can then be eliminated by quoting the Axiom of Replacement a finite number of times.

7. HIGHER AXIOMS AND MODELS FOR SET THEORY

In the discussion of models in Chapter I we used intuitive set theory. It is now clear that all that discussion can be easily formalized within ZF. Although ZF does not contain symbols for arbitrary formal systems, by assigning numbers (or even perhaps ordinals) to the symbols used one can, as already explained, "arithmetize" proofs and stay entirely within ZF. Thus, in ZF we can state the following axiom:

Axiom M.

There is a set M and a binary relation $\epsilon \subseteq M \times M$ which makes M a model for ZF.

We emphasize that Axiom M is a single statement in ZF, despite the fact that the axioms of ZF are infinite. This is because the rule for their formation is expressible as a single statement. The Completeness Theorem now says that Axiom M is equivalent to Consis ZF and thus this axiom does not lead to any new point of view. One can, however, also state

Axiom SM.

There is a set M such that if R is $\{\langle x,y\rangle \mid x \in y \ \& \ x \in M \ \& \ y \in M\}$ then M is a model for ZF under the relation R.

The model M is called a standard model since the ϵ-relation used is merely the standard ϵ-relation. Both Axioms M and SM are unprovable in ZF since they imply Consis ZF. On the other hand, both are most probably "true" and could be considered as axioms. Certainly, Axiom M is true since if one believes in sets, one certainly believes that no contradiction exists in ZF. To see informally that SM is "true" we proceed as follows:

The Löwenheim-Skolem theorem allows us to pass to countable sub-models of a given model. Now, the "universe" does not form a set and so we cannot, in ZF, prove the existence of a countable sub-model. However, informally we can repeat the proof of the theorem. We recall that the proof merely consisted of choosing successively sets which satisfied certain properties, if such a set existed. In ZF we can do this process finitely often. There is no reason to believe that in the real world this process cannot be done countably many times and thus yield finally a countable standard model for ZF. The only reason this cannot be done in ZF is simply that there is no property $A(n,x)$ in ZF which expresses, for each n, the property of x which we wish to consider at the nth stage.

We shall later show that Axiom M does not imply Axiom SM.

These axioms are examples of how we can naturally extend the axioms of ZF by adjoining intuitively true statements. The Incompleteness Theorem tells us that one way of extending an axiom system is to adjoin the number-theoretic statement that the system is consistent. If carried on as a transfinite process this gives an unlimited supply of new axioms. However, there are more powerful axioms of a more set-theoretic character that can also be adjoined. These are the so-called Higher Axioms of Infinity. We know that any particular integer can be shown to exist by using the axiom of pairs. The usual axiom of infinity makes a great leap by assuming the existence of one set which contains all the integers. That is, it asserts the existence of a large set which cannot be obtained by the repetition of a certain process. The following axiom is also of this character.

<u>Axiom of inaccessible cardinals</u>.

There is an uncountable cardinal A such that

1) if $B < A$, A is <u>not</u> the sum of B cardinals each less than A.
2) if $B < A$, then $\overline{\overline{P(B)}} < A$.

Thus A cannot be reached by using the power set operation applied to lower cardinals, nor by using the operation given to us by the Axiom of Replacement, that is, taking the range of a function, when applied to lower cardinals. It is clear that if A is an inaccessible cardinal, then the set of all sets of rank less than A is a model for ZF, since the only two troublesome axioms, Power Set and Replacement, do not lead out of the set of cardinals less than A. Since this proves Consis ZF, the incompleteness theorem implies that this axiom is unprovable in ZF. The general tendency is to accept the axiom as true or at least that it is consistent with ZF. One is inclined to be generous and say "Why can't such a large set exist?" To say that it is false would seem to put an unjustified bound on the size of cardinals. To someone committed to the idealist position that infinite sets really "exist", the axiom is either true or false. For others it is only a question of whether or not it is consistent. It is possible to postulate the existence of two inaccessible cardinals and to repeat the process in still more complicated ways. A systematic study of these axioms of infinity was made by Mahlo and more recently by A. Lévy. Of course, no single scheme for generating axioms is ever complete since one can always go one step further.

In general, each new axiom will allow one to prove the consistency of a weaker system. Thus each axiom will prove a statement in elementary number theory not previously provable. To dismiss these axioms as irrelevant is thus to essentially give up any hope of proving these statements. It is also, however, quite difficult when confronted with an axiom of infinity to decide whether to accept it, since as one postulates the existence of larger sets one also comes closer to the paradoxes arising from the set of all sets, etc. Later, we shall show that the Continuum Hypothesis poses an even more disturbing problem in that it remains unprovable even after any higher axiom of infinity, as we presently understand the term, is added.

There is one particular axiom which deserves special attention because it seems in some ways to be an axiom of infinity, yet is more powerful than any introduced up to now. Soon after measure theory was discovered, the problem arose to find a set S and a countably additive non-trivial finite real-valued measure defined on <u>all</u> subsets of S such that each point has measure zero. It is well-known that ordinary Lebesgue measure on the real line does not have this property.

A set S which has such a measure is called a measurable cardinal. It was first shown that if such a set S exists its cardinality must exceed $\aleph_1, \aleph_2, \ldots$ up to quite high cardinals. More precisely, if one assumes the Generalized Continuum Hypothesis one can show that the cardinality of S must exceed the first inaccessible cardinal. Recently it was shown that even the first, second, etc. inaccessible cardinals do not support such measures and it now appears probable that, loosely speaking any cardinal which is defined as the first cardinal satisfying a given axiom of infinity is not measurable. An important feature of measurable cardinals is that their existence contradicts the Axiom of Constructibility (to be defined in Chapter III), which is not the case for the usual axioms of infinity. The above results seem to suggest that in some sense postulating the existence of a measurable cardinal asserts only the existence of a very large set but the intuition is not at all clear. Probably few people expect to disprove the existence of a measurable cardinal.

There is another method to extend the axioms and that is to allow sets corresponding to more complicated properties. Thus the set $\{n \mid A_n$ is true$\}$ where A_n is the nth statement of ZF might be admitted. This requires a new notation and is best thought of in GB. That is one can adjoin an axiom scheme which says, if $A(u, Y_1, \ldots, Y_n)$ is a formula with one set variable, n class variables and which may involve <u>bound class variables</u>, then

$$\forall X_1, \ldots, X_n \ \exists Y (u \in Y \leftrightarrow A(u, X_1, \ldots, X_n)) \ .$$

This will allow us to define classes by induction and for example to define the "truth set" given above. One can go further by allowing as a new primitive, super-classes" whose members are classes. New axioms of comprehension, can then be added allowing us to form super-classes corresponding to different formulas. There is certainly a strong relation between these higher comprehension axioms and higher axioms of infinity. Although one type of axiom system in general does not imply the other, the second type is stronger in that it usually implies the consistency of a corresponding system of the first kind. This is done by considering as the model the set of all sets of rank less than a very high cardinal.

8. LÖWENHEIM-SKOLEM THEOREM REVISITED

In the last section we spoke of how a loose application of the Löwenheim-Skolem theorem proves the existence of a countable standard model. That argument of course cannot be formalized in ZF. Here we shall prove, in ZF, that if one is only concerned with finitely many statements holding in the model, the result can be proved. By taking conjunctions we can restrict our attention to a single statement. This result will play an important role in the next chapter.

THEOREM. Let $A(x_1,\ldots,x_n)$ be a formula in ZF. One can prove in ZF that for any set S there is a set $S' \supseteq S$ such that $\bar{\bar{S}}' = \max(\aleph_0, \bar{\bar{S}})$ and for all $\bar{x}_i \in S'$, $A(\bar{x}_1,\ldots,\bar{x}_n) \leftrightarrow A_{S'}(\bar{x}_1,\ldots,\bar{x}_n)$ where $A_{S'}$ is A with all quantifiers restricted to S'.

Proof. We follow the proof of the original theorem but give our proof entirely in ZF. Assume A is of the form $Q_1 y_1,\ldots,Q_m y_m$ $B(x_1,\ldots,x_n,y_1,\ldots,y_m)$ and B has no quantifiers. For any set T, we claim that there are functions $f_r(\bar{x}_1,\ldots,\bar{x}_n,\bar{y}_1,\ldots,\bar{y}_{r-1})$ defined for $\bar{x}_i$ and $\bar{y}_j$ in T with the following property: If $Q_r = \exists$ and there is a set $\bar{y}_r$ such that

$$(1) \qquad Q_{r+1}y_{r+1},\ldots,Q_m y_m \, B(\bar{x}_1,\ldots,\bar{x}_n,\bar{y}_1,\ldots,\bar{y}_r.y_{r+1},\ldots,y_m)$$

then (1) holds with $\bar{y}_r = f_r(\bar{x}_1,\ldots,\bar{x}_n,\bar{y}_1,\ldots,\bar{y}_{r-1})$. If no $\bar{y}_{r+1}$ exists satisfying (1), then $f_r = \emptyset$. If $Q_r = \forall$, then f_r has the same property with (1) replaced by its negation. The proof that f_r exist uses the Axiom of Choice. However, since the Axiom of Choice is only stated for sets (whereas here we require it for a class in the sense of GB), we proceed as follows. For $\bar{x}_1,\ldots,\bar{x}_n,\bar{y}_1,\ldots,\bar{y}_{r-1}$ all in T let $R(\bar{x}_i,\bar{y}_j)$ be the <u>set</u> of all $\bar{y}_r$ of least rank satisfying (1) if $Q_r = \exists$ or its negation if $Q_r = \forall$. $R(\bar{x}_i,\bar{y}_j)$ is a set since it is a subset of the set of all sets of some rank. By the Axiom of Choice we can pick one element from $R(\bar{x}_i,\bar{y}_j)$ to obtain f_r. Let T* be the union of T and the range of all the f_r. By another application of the Axiom of Choice it follows that we can construct a sequence of sets S_n with

$S_o = S$ and $S_{n+1} = S_n^*$ for some functions f_r satisfying our property. Put $S' = \bigcup_n S_n$. Clearly $\bar{\bar{S}}' = \max(\aleph_o, \bar{\bar{S}})$. The usual proof shows that $A(\bar{x}_1, \ldots, \bar{x}_n) \leftrightarrow A_{S'}(\bar{x}_1, \ldots, \bar{x}_n)$. For, let $C(\bar{x}_1, \ldots, \bar{x}_n, \bar{y}_1, \ldots, \bar{y}_r)$ denote the statement (1). Assume that we have shown that $r > r_o$, $C(\bar{x}_1, \ldots, \bar{x}_n, \bar{y}_1, \ldots, \bar{y}_r) \leftrightarrow C_{S'}(\bar{x}_1, \ldots, \bar{x}_n, \bar{y}_1, \ldots, \bar{y}_r)$ for all $\bar{x}_i$ and $\bar{y}_j$ in S'. Certainly this is the case if $r_o = m$. Then with $r = r_o$, given $\bar{x}_i$, $\bar{y}_j$ in S' they must all lie in S_k for some k. $Q_{k+1} = \exists$, then if $C(\bar{x}_i, \bar{y}_j)$ is true $f_{r+1}(\bar{x}_i, \bar{y}_j)$ is in S_{k+1} and thus by assumption $C(\bar{x}_1, \ldots, \bar{x}_n, \bar{y}_1, \ldots, \bar{y}_{r+1})$ is true in S' so $C(\bar{x}_1, \ldots, \bar{x}_n, \bar{y}_1, \ldots, \bar{y}_r)$ is true in S'. If $C(\bar{x}_i, \bar{y}_j)$ is false then for any $\bar{y}_{r+1}$, $C(\bar{x}_1, \ldots, \bar{x}_n, \bar{y}_1, \ldots, \bar{y}_{r+1})$ is false and hence false in S' so $C(\bar{x}_1, \ldots, \bar{x}_n, \bar{y}_1, \ldots, \bar{y}_r)$ is false in S'. A similar argument holds if $Q_{r+1} = \forall$ and the theorem is proved.

COROLLARY. No finite number of axioms of ZF imply all the axioms of ZF.

<u>Proof</u>. Let (ZF)' be a given finite collection of axioms of ZF such that $(ZF)' \rightarrow ZF$. Then (ZF)' is powerful enough so that the Incompleteness Theorem applies and so Consis(ZF)' is unprovable in (ZF)'. However, we have just seen that Consis(ZF)' is provable in ZF and hence by our assumption also in (ZF)' which is a contradiction.

We give a trivial theorem characterizing standard models.

THEOREM. Let M be a model for ZF, and let xRy denote the relation between x and y expressing that x is a member of y in the model M. A necessary and sufficient condition for M to be isomorphic to a standard model is that there does not exist a sequence $\langle x_n \rangle$ of elements in M (the sequence itself need not be in M) such that $x_{n+1} R x_n$.

<u>Proof</u>. In the proof, ordinal, function, etc. will refer to the "actual" notions and not to the corresponding notions in M. Define by transfinite induction on α the sets S_α and functions f_α on S_α as

follows. S_0 consists only of $\emptyset_M$, the empty set in M (which is unique since extensionality holds in M) and $f_0(\emptyset_M) = \emptyset$. Given S_β and f_β for $\beta < \alpha$, let S_α be those elements x in M such that $yRx \to y \in S_\beta$ for some $\beta < \alpha$, and such that x is not in any S_β for $\beta < \alpha$. Let $f_\alpha(x) = \{f_\beta(y) | \beta < \alpha, y \in S_\beta, yRx\}$ for $x \in S_\alpha$. We claim that if α_0 is the first ordinal such that $S_{\alpha_0+1} = \emptyset$, then $\cup\{S_\beta | \beta \leq \alpha_0\} = M$. For if we put $T = \cup\{S_\beta | \beta \leq \alpha\}$ and if $x \in M\text{-}T$, then since $x \neq \emptyset_M$ there is a y_1 such that y_1Rx. Since $x \notin T$, we have $y_1 \notin T$. Similarly, there is a y_2 such that y_2Ry_1, $y_2 \notin T$, etc., so that we obtain a sequence y_n, $y_{n+1}Ry_n$, a contradiction. It is left as an exercise for the reader to show that $f = \cup\{f_\alpha | \alpha \leq \alpha_0\}$ is a function which maps M isomorphically onto the range of f.

CHAPTER III

THE CONSISTENCY OF THE CONTINUUM HYPOTHESIS AND THE AXIOM OF CHOICE

1. INTRODUCTION

Throughout this chapter we do not include the Axiom of Choice (AC) as one of the axioms of ZF. Our goal is to present Gödel's proof that if ZF is consistent then it remains consistent if the Generalized Continuum Hypothesis (GCH) and AC are added. The Continuum Hypothesis (CH) was first stated by Cantor in 1878. It was listed by Hilbert first in his list of unsolved problems given in his famous address of 1900. Despite many attempts the problem remained unsolved. It was however used freely in proofs since it often simplified situations. In 1938 Gödel [12] proved his consistency result. Although CH played an insignificant role in the development of mathematics, AC had been used extensively and hence Gödel's theorem was very reassuring. We shall now give some motivation for the method he used.

If CH is true then the real numbers can be put in one-one correspondence with $\aleph_1$, the set of all countable ordinals. If we think of an ordinal as representing a "process" we will thus have a method of assigning to each countable "process" a unique real number. Thus, it is not unreasonable to imagine a transfinite process of constructing real numbers, or even sets in general, which is sufficiently general so as to possibly include all sets. Now various philosophical viewpoints have been presented which would only accept those mathematical objects which had been effectively constructed. The strictest of these would only accept finite sets and thus clearly has no relevance for set theory. The correct notion, for this purpose, is that of <u>predicative definitions</u>. Let us take the point of view that a set is not an a priori given object but must be defined by a property. Since ZF allows such highly non-constructive arguments we shall be liberal and allow any reasonable property

$P(x)$ even if it can in no way be empirically verified for a given x. However, it does seem reasonable to demand that whatever $P(x)$ is, it speaks only about objects which have previously been admitted. For example, suppose we accept the integers as given as well as the arithmetic operations. Then we might consider a set of integers defined by a property $P(n)$ which has all its bound variables ranging only over integers. Such a set of integers is said to be predicatively defined in terms of the integers. Consider however the following property. An integer n satisfies $P(n)$ if there is a partition of the set ω of integers into n disjoint sets none of which contains arithmetic progressions of arbitrary length. Let S be $\{n \mid P(n)\}$. In order to determine whether e.g. $5 \in S$, we have to consider <u>all</u> partitions of ω into 5 sets including possibly partitions in which the set S itself occurs. Thus, in order to define the one set S we are required to accept as a meaningful totality the set of all sets of integers. This is an example of an impredicative definition. From the point of view of ZF this definition is perfectly acceptable. However from a more constructive point of view the definition leaves something to be desired. The important point here is not that the definition cannot be effectively carried out, but that even the domain of discourse has not been sufficiently specified. Certainly this is the kind of situation (however presumably here in an innocuous form), which occurs in the paradoxes. Attempts were made therefore to develop set theory on a predicative basis. In Principia Mathematica, Russell and Whitehead gave a development of "type theory" which avoided certain aspects of impredicativity. Also H. Weyl in his book "The Continuum" showed how one could develop most, but not all, of classical analysis along predicative lines. Of course, the impredicative character of the Replacement Axiom is an essential feature of ZF and no strictly predicative development can be truly relevant for ZF. However, we are searching for a correspondence between ordinals and sets, and it was Gödel's idea that if we iterate predicative constructions up to any ordinal the resulting sets could furnish an adequate model for ZF. Philosophically we will have gained nothing since we do not say how our ordinals are obtained nor do we distinguish between "predicative" and "impredicative" ordinals. Since we are here concerned with a relative consistency question, however, this notion is extremely useful. A set which can be obtained as the result of a transfinite sequence of predicative definitions Gödel called "constructible". His result then is that the constructible sets are a model for ZF and that in this model GCH and AC hold. The notion of a predicative construction must be made more precise, of course, but there is essentially only one way to proceed. Another way to explain constructibility is to remark that the constructible sets are those sets which must occur in any model in which one admits all ordinals.

The definition we now give is the one used [12].

<u>Definition</u>. Let X be a set. The set X' is defined as the union of X and the set of all sets y for which there is a formula $A(z,t_1,\ldots,t_k)$ in ZF such that if A_X denotes A with all bound variables restricted to X, then for some $\bar{t}_i$ in X,

$$y = \{z \in X \mid A_X(z,\bar{t}_1,\ldots,\bar{t}_k)\}$$

Observe $X' \subseteq P(X) \cup X$, $\bar{\bar{X}}' = \bar{\bar{X}}$ if X is infinite (and we assume AC). It should be clear to the reader that the definition of X', as we have given it, can be done entirely within ZF and that $Y = X'$ is a single formula $A(X,Y)$ in ZF. In general, one's intuition is that all normal definitions can be expressed in ZF, except possibly those which involve discussing the truth or falsity of an infinite sequence of statements. Since this is a very important point we shall give a rigorous proof in a later section that the construction of X' is expressible in ZF. However, it is an excellent exercise for the reader at this point to give the definition himself.

<u>Definition</u>. If On α, we define M_α by $M_0 = \emptyset$ and $M_\alpha = (\bigcup_{\beta<\alpha} M_\beta)'$.

<u>Definition</u>. A set x is constructible if $\exists\alpha$, On α and $x \in M_\alpha$.

Since $X' \subseteq P(X) \cup X$ it is clear that all constructible sets are well-founded and if $x \in M_\alpha$, rank $x \leq \alpha$. This construction is clearly related to the definition of well-founded sets and to the consistency proof for the Regularity Axiom. We shall denote the "class" of constructible sets by L, so that $x \in L$ means merely that x is constructible (and does not mean that L is a set) and for any formula A, A_L denotes the same formula with all variables restricted to be constructible. Since the universal class was denoted by V in [14], the statement that every set is constructible, which is known as the Axiom of Constructibility is sometimes written $V = L$.

Since our definition of M_α proceeds by a rather simple induction it is clear that it can be formalized in ZF. Thus the statement $V = L$ is a single statement in ZF. The main results now are the following:

THEOREM 1. If A is any axiom of ZF, then A_L is provable in ZF.

This means intuitively that the constructible sets are a model for ZF. We cannot say it in this form since L is not a set.

THEOREM 2. $(V = L)_L$ is provable in ZF.

This is a small but subtle point. It says that a constructible set is constructible when the whole construction is relativized to L.

THEOREM 3. $(V = L) \rightarrow$ AC & GCH is provable in ZF.

That is, the Axiom of Constructibility implies AC and GCH so that taken together the theorems show that AC and GCH cannot be disproved in ZF.

Before proceeding, we should like to clarify the relationship of the exposition of [12] to that of [14]. The approach of [12] which we have just outlined is probably more easily understandable than that of [14]. Of course the two approaches are substantially identical. However because of the widespread influence of [14] we shall later also give that approach and show the equivalence of the two methods.

2. PROOF OF THEOREM 1

In this section we consider Theorem 1, namely that the axioms hold in L. Some of the axioms are handled quite trivially.

1. Since $\emptyset \in L$, the Axiom of the Null Set holds in L.

2. Since $X' \subseteq P(X) \cup X$ for all X, it follows that if $x \in M_\alpha$ and $y \in x$ then $y \in \bigcup_{\beta < \alpha} M_\beta$ so $y \in L$. Thus each M_α is transitive and so if $x \in L$, $y \in x \to y \in L$. The Axiom of Extensionality therefore holds in L.

3. Assume $x \in M_\alpha$, $y \in M_\beta$, $\alpha \leq \beta$. Then x and y are both in M_β. Now, $\{x,y\} = \{z | z \in M_\beta \ \& \ (z = x \vee z = y)\}$ so that $\{x,y\} \in M_{\beta+1}$. Thus, the Axiom of Unordered Pairs holds in L.

4. If $x \in M_\alpha$, the set $\{z | z \in M_\alpha \ \& \ \exists y(y \in M_\alpha \ \& \ y \in x \ \& \ z \in y)\}$ is in $M_{\alpha+1}$. But since M_α is transitive this is exactly the sum-set of x so that the Sum-Set Axiom holds in L.

5. To prove the Axiom of Infinity we first state two lemmas that shall be useful later.

LEMMA. Let $A(x)$ be the formula "x is an ordinal". Then $\forall x (x \in L \to (A(x) \leftrightarrow A_L(x)))$. In general, if X is any transitive set or class, $A(x) \leftrightarrow A_X(x)$.

<u>Proof</u>. We are assuming Regularity, so the condition that x is an ordinal is a condition solely on the elements of x. Since L is transitive this condition is therefore the same when relativized to L.

LEMMA. For all ordinals α, $\alpha \in M_{\alpha+1}$.

<u>Proof</u>. Clearly $\emptyset \in M_1$. Let α be the least α for which the lemma is false. Then $\beta < \alpha \to \beta \in M_{\beta+1} \to \beta \in X = \bigcup_{\gamma \leq \alpha} M_\gamma$. Let $A(x)$ be the formula "x is an ordinal". Since X is transitive and by the above lemma we have that $x \in X \to (A_X(x) \leftrightarrow A(x))$. The set $\gamma = \{x \in X | A_X(x)\}$ is thus a set of ordinals and by the transitivity of X is actually an ordinal itself which exceeds all the ordinals $\beta < \alpha$ and hence $\gamma \geq \alpha$. By the transitivity of X', $\alpha \in X' \subseteq M_{\alpha+1}$.

Since ω thus is in L, we have proved the Axiom of Infinity.

6. The Axiom of Regularity is trivial and is left as an exercise.

7. The first non-trivial axiom to consider is the Power Set Axiom. Let $x \in L$. Let $P(x)$ be the power set of x and $P_L(x) = \{y \mid y \in P(X) \; \& \; y \in L\}$. For each $y \in P_L(x)$ let $\varphi(y)$ = least α such that $y \in M_\alpha$. By the Replacement Axiom there is an ordinal β which is the sup of the ordinals in the range of φ. Hence $y \in P_L(x) \rightarrow y \in M_\beta$. Consider the set $\{y \mid y \in M_\beta \; \& \; \forall t(t \in M_\beta \rightarrow (t \in y \rightarrow t \in x))\}$. This set is in M'_β and is clearly equal to $P_L(x)$. Thus $P_L(x)$ is constructible which implies that the Power Set Axiom holds in L.

Observe that if $x \in M_\alpha$ this proof does not explicitly yield a β such that $P_L(x) \in M_\beta$. We shall later give a precise value for β in terms of α. Of course, we have used the Power Set Axioms to prove that the corresponding relativized axiom also holds. However, it is perhaps more surprising that we have also used the Replacement Axiom. This is somewhat unesthetic and also puts the Replacement Axiom in too central a position. For example, if we were to work only with the Separation Axiom we would be unable to give this proof. There is another proof which we shall give later when we discuss GCH in L that avoids Replacement and which also gives a more effective bound for the ordinal that constructs the power set. With this proof one can roughly say that to prove the Power Set Axiom in L requires essentially only the Power Set Axiom together with a very weak form of a Replacement or Comprehension Axiom.

8. The Replacement Axiom is the most difficult to verify in L. Let $A(x,y;t_1,\ldots,t_n)$ be a formula in ZF and let A_L denote the relativized formula. If $t_i \in L$ and A_L defines $y = \varphi(x)$ as a univalent function in L, we have to show that if $u \in L$, the range v of φ on u is in L.

LEMMA. Let $y = \varphi(x)$ be a univalent function defined by a formula $A(x,y;t_1,\ldots,t_n)$ for some t_i and such that $x \in L$ implies $\varphi(x) \in L$. If $u \in L$ then $\exists w \in L$ such that if v is the range of φ on u, then $v \subseteq w$.

Proof. For each x in u let $g(x)$ be the least α such that $\varphi(x) \in M_\alpha$. Let $\beta = \sup\{g(x) | x \in u\}$. Clearly $v \subseteq M_\beta$ and $M_\beta \in L$.

To proceed further we need the Löwenheim-Skolem theorem of Chapter II, § 8 but in a slightly modified form to suit our purposes. In Chapter II, we cut down from the entire universe V to a set, whereas here we cut down from L to a set. Also we need to know that the set is in L.

THEOREM. Let $A(x_1,\ldots,x_n)$ be a formula in ZF in which all variables, bound and free, are restricted to be in L (i.e., A is already relativized to L). Let $S \in L$. There exists $S' \in L$ such that $S' \supseteq S$ and for all $\bar{x}_i \in S'$, $A(\bar{x}_1,\ldots,\bar{x}_n) \longleftrightarrow A_{S'}(\bar{x}_1,\ldots,\bar{x}_n)$.

Proof. We mean of course that for each A, the above is a theorem in ZF. We follow the proof of Chapter II, § 8. Let A be of the form $Q_1y_1 \cdots Q_my_m\, B(x_1,\ldots,x_n,y_1,\ldots,y_m)$. Let $T \in L$. For $1 \leq r \leq m$ there are functions $f_r(\bar{x}_1,\ldots,\bar{x}_n,\bar{y}_1,\ldots,\bar{y}_{r-1})$ defined for $\bar{x}_i$, $\bar{y}_j$ in T with the following property: If $Q_r = \exists$ and there is a set $\bar{y}_r$ such that

(1) $$Q_{r+1}y_{r+1} \cdots Q_my_m\, B(\bar{x}_1,\ldots,\bar{x}_n,\bar{y}_1,\ldots,\bar{y}_r, y_{r+1},\ldots,y_m)$$

then $f_r = \alpha$ where α is the least ordinal such that there is a $\bar{y}_r \in M_\alpha$ satisfying (1). Because of the restriction on A if a $\bar{y}_r$ exists satisfying (1) it must be in L so that f_r is well defined. If no such $\bar{y}_r$ exists, put $f_r = 0$. If $Q_r = \forall$, then f_r is defined the same way with (1) replaced by its negation. Let β be the supremum of $f_r(\bar{x}_i,\bar{y}_j)$ for all $\bar{x}_i$ and $\bar{y}_j$ in T and all r, $1 \leq r \leq m$. Put $T^* = T \cup M_\beta$. Now, we define a sequence S_n with $S_0 = M_\alpha$ if α is the least ordinal such that $S \in M_\alpha$, and $S_{n+1} = S_n^*$. The same proof as before now shows that $S' = \bigcup_n S_n$ has the required property of the theorem. Also S' clearly equals $\bigcup\{M_\beta | \beta < \alpha\}$ for some α and this set is clearly in M_α. We make no cardinality statement about S', since we do not need it and AC is not available.

We can now prove the Replacement Axiom in L. Let $A(x,y;t_1,\ldots,t_n)$ be a statement which is relativized to L and such that for particular

$t_1, \ldots, t_n$ defines $y = \varphi(x)$ as a univalent function in L. Let $u \in L$ and let v be the range of φ on u. By our lemma, there is an α such that $v \subseteq M_\alpha$ and we may clearly assume that $u, t_1, \ldots, t_n$ also belong to M_α. Taking M_α as the S of the theorem, it follows that for some S' in L, $A(x,y;t_1,\ldots,t_n) \longleftrightarrow A_{S'}(x,y;t_1,\ldots,t_n)$ for all x, y in S'. We know $S' \in M_\beta$ for some β so that since $v = \{y \in S' \mid \exists x \in u \,\&\, A_{S'}(x,y;t_1,\ldots,t_n)\}$, v is defined by a condition in which all variables are restricted to M_β and then $v \in (M_\beta)' = M_{\beta+1}$ so $v \in L$ and the Replacement Axiom holds. Theorem 1 is thus completely proved.

3. ABSOLUTENESS

Before we proceed to prove Theorem 2, we shall show that the relation $Y = X'$ can be expressed in ZF, and that consequently the statement $V = L$ can also be expressed. For each $r \geq 0$ let X_r denote the set of all sets S of n-tuples $\langle x_1, \ldots, x_n \rangle$ for which there is a formula $A(x_1, \ldots, x_n; t_1, \ldots, t_m)$ with exactly r quantifiers, and $\bar{t}_i \in X$ such that $S = \{\langle x_1, \ldots, x_n \rangle \mid A_X(x_1, \ldots, x_n; \bar{t}_1, \ldots, \bar{t}_m)\}$. We show that the relation $Y = X_r$ is expressible in ZF. What is essentially involved here is the analysis of formulas into their atomic pieces in precisely the manner as was done in showing that the class axioms of GB were sufficient to prove the Replacement Axiom of ZF. The relation $Y = X_0$ is expressed by enumerating all the formulas with no quantifiers and by induction on the length of the formulas, using the various Boolean operations, defining the sets S arising from each formula. By induction on r, one now defines $Y = X_r$, by saying that a set S of n-tuples belongs to X_r, if there is a set T of (n+1)-tuples in X_{r-1} such that $\langle x_1, \ldots, x_n \rangle \in S \longleftrightarrow \exists x_0 \in X$ such that $\langle x_0, x_1, \ldots, x_n \rangle \in T$ or if there is a set T of (n+1)-tuples in X_{r-1} such that $\langle x_1, \ldots, x_n \rangle \in S \longleftrightarrow \forall x_0 \in X, \langle x_0, \ldots, x_n \rangle \in T$. Then X' is finally defined as the union of X and the set of all sets of 1-tuples which occur in any X_r. Since the sets M_α were defined by a simple transfinite induction in terms of the operation $X \to X'$, it is clear that $V = L$ is thus expressed as a statement in ZF.

To prove Theorem 2 we have to show that the construction of the M_α has the same result when carried out in the model L as when carried out in the universe L. Now, the fact that there are sets M_α in L satisfying the relativized definition of M_α follows from the fact that L is a model for ZF and that thus the theorem of transfinite induction and the existence of the operation $X \to X'$ hold in L. The fact that the M_α so constructed are the same as the original M_α and hence that $(V = L)_L$ holds is a consequence of a very general fact and does not depend on the fact that L is a model for ZF. Namely, if we are operating within a given transitive set or class B in which the construction of the M_α can be performed relativized to B for a given α in B (because B contains all the sets it need have) then the resulting M_α must be the same as the ordinary M_α. This is because of the predicative character of the construction which implies that to check that $x \in M_\alpha$ requires only looking at all M_β with $\beta < \alpha$ and is not at all affected by the existence or absence of any other sets in the universe. We now do this in a formal manner.

Let $B(x)$ denote any formula which may involve other free variables which we regard as fixed. In our main application $B(x)$ will denote "$x \in L$". Assume $B(x)$ is transitive, i.e., that $B(x) \,\&\, y \in x \to B(y)$. For any formula A we denote by A_B the formula obtained by demanding that all the variables in A satisfy $B(x)$. Since for fixed t we can take $B(x) \equiv x \in t$, we see that relativization to a set is a particular case of this type of relativization.

<u>Definition</u>. A formula $A(x_1,\ldots,x_n)$ is said to define an <u>absolute</u> relation if for all transitive conditions $B(x)$, we have $B(x_1) \,\&\, \cdots \,\&\, B(x_n) \,\&\, A_B(x_1,\ldots,x_n) \to A(x_1,\ldots,x_n)$. Furthermore, if B' is another condition such that $\forall x\, B(x) \to B'(x)$, then $B(x_1) \,\&\, \cdots \,\&\, B(x_n) \,\&\, A_B(x_1,\ldots,x_n) \to A_{B'}(x_1,\ldots,x_n)$.

Intuitively this means that to check $A(x_1,\ldots,x_n)$ it is enough to check it in any transitive class which is large enough to contain all the objects that A speaks of. It would be absurd to demand that $A \to A_B$ since we have no assurance that the class of B will be big enough. One could introduce a stricter notion of absoluteness which says that A is absolute if whenever A is relativized to a class B, for which the axioms of ZF hold when relativized to B, then $A \leftrightarrow A_B$. This is indeed the case for the conditions we are interested in. The term absolute is used in the literature with different meanings and the reader should carefully note our use of the term.

Let us begin by giving examples of relations which are not absolute. The relation $y = P(x)$ is not absolute since y may be the set of all subsets of x in the class B yet not be the true power set. Here we see the impredicative character of the Power Set Axiom which requires searching the entire universe for all possible subsets. Another example is the relation $\bar{\bar{x}} < \bar{\bar{y}}$. If there is no one-one mapping of x onto y in a class B this does not imply that there can not exist such a mapping.

We now give a sequence of absolute relations. The absoluteness of each is obvious if one takes into account the absoluteness of its predecessors. The transitivity of B plays a role in some e.g., the relation "$z = \langle x,y\rangle$" since if z had members not in B we could not test whether z is an ordered pair by remaining in B.

1. $x \in y$
2. $x = y$
3. $x \subseteq y$
4. $x \subset y$
5. $x = \emptyset$
6. $z = x \cap y$
7. $z = \{x,y\}$
8. $z = \langle x,y\rangle$
9. z is an ordered pair
10. $z = x \times y$
11. x is a function
12. x is an ordinal
13. $x = \omega$ (i.e., satisfies the definition of ω)
14. $Y = X_r$ where X_r has been defined in this section
15. $Y = X'$
16. $x = M_\alpha$
17. α is an ordinal and $x = M_\alpha$

THEOREM. $(V = L)_L$.

Proof. We proved the existence of a set M_α for each α satisfying a certain property. This proof used only the axioms of ZF and since these hold in L, there is a set M_α in L satisfying the relativized condition. Now, if $x \in L$ there is an α such that $x \in M_\alpha$ and we also have seen that $\alpha \in L$. But the above results on absoluteness imply that the set M_α defined relative to L is still M_α so that $x \in M_\alpha$ holds when relativized to L.

4. PROOF OF AC AND GCH IN L

Since the Axiom of Constructibility, $V = L$, holds in L to show that AC and GCH hold in L it suffices to prove Theorem 3, namely, that $V = L \rightarrow \mathrm{AC}\ \&\ \mathrm{GCH}$

THEOREM. There is a formula $A(u,v,X,Y)$ in ZF, such that if Y is a well ordering of the set X, the relation $u < v \leftrightarrow A(u,v,X,Y)$ induces a well-ordering of the set X'.

Proof. Enumerate the countably many formulas $B_n(x; t_1, \ldots, t_k)$. We have already essentially shown how to express in ZF the relation $C(u,n,t_1,\ldots,t_k)$, $u = \{x \in X | B_{n,X}(x; t_1, \ldots, t_k)\}$. Now the well-ordering Y induces a natural well-ordering on the set of all possible $(k+1)$-tuples $\langle n, t_1, \ldots, t_k \rangle$ where $t_i \in X$. For each $u \in X'$, we can define $\varphi(u)$ as the first $(k+1)$-tuple, for some k, under this well-ordering, such that $C(u, \varphi(u))$ holds. Now we can define A by having $u < v$ mean $\varphi(u) < \varphi(v)$.

By transfinite induction, we can define a well ordering on M_α as follows. If the well-ordering has been defined for all M_β with $\beta < \alpha$, we well-order $\bigcup_{\beta < \alpha} M_\beta$ in an obvious manner. Then our theorem allows us to define a well-ordering on M_α. Thus if $V = L$ holds, let $\varphi(x)$ be the least α such that $x \in M_\alpha$. Define $x < y$ if $\varphi(x) < \varphi(y)$ or if $\varphi(x) = \varphi(y) = \alpha$ and x precedes y in the well-ordering of M_α. Thus we have given a single formula $A(x,y)$ which well-orders <u>all</u> sets. In the language of GB we have a class which well-orders the universe. Thus $V = L \rightarrow \mathrm{AC}$.

The proof that $V = L \rightarrow \mathrm{GCH}$ is the most technically difficult point of Gödel's proof and was the traditional stumbling block for readers of [14]. Since we know $V = L \rightarrow \mathrm{AC}$ we shall not hesitate to use AC. From AC it easily follows that if X is infinite, $\bar{\bar{X}}' = \bar{\bar{X}}$. An obvious transfinite induction now shows that for infinite α, $\bar{\bar{M}}_\alpha = \bar{\bar{\alpha}}$. We shall use $V = L$ without explicit mention.

<u>THEOREM 1</u>. If $x \in M_\alpha$, α infinite and $y \subseteq x$ then $\exists \beta$, $\bar{\bar{\beta}} = \bar{\bar{\alpha}}$ and $y \in M_\beta$.

This theorem is the heart of the matter. Since $\omega \in M_\omega$ it will follow that every set of integers is constructed after countably many steps. Since the number of countable ordinals is $\aleph_1$ and $\bar{\bar{M}}_\alpha = \aleph_0$ if α is countable, this would clearly imply that the continuum $C = \aleph_1$. Let us attempt to give some intuitive justification for why all sets of integers are constructible by countable ordinals. If $x \subseteq \omega$, $x \in M_\alpha$ then one can ask what are the essential properties of α which imply that $x \in M_\alpha$. Now x is determined by the truth value of the countably many statements "$n \in x$". For each n we can think of this as imposing one condition on α. Thus it is not unreasonable that these countably many conditions, if they can be satisfied by any α, can also be satisfied by a countable α. The mechanism for making this precise will be furnished by the Löwenheim-Skolem theorem which allows us to construct smaller sets having the same properties as larger sets. Also an important role will be played by the trivial result of Chapter II, § 6 concerning $\in$-isomorphisms.

The proof of the above theorem will also furnish us with another method by which we could have proved the Power Set Axiom. Namely, if $x \in M_\alpha$, clearly $P(x) \in M_\beta$ where β is the first ordinal of cardinality greater than α. This is the proof we mentioned in § 2.

Let α now be infinite $x \in M_\alpha$, $y \subseteq x$ and $y \in M_\beta$. Recall that a set T is called extensional if $x, y \in T$ and $x \neq y$ implies $\exists z \in T \; (z \in x \;\&\; \sim z \in y) \vee (z \in y \;\&\; \sim z \in x)$.

LEMMA 1. There is an extensional set T, $\bar{\bar{T}} = \bar{\bar{\alpha}}$, such that $M_\alpha \subseteq T$ and M_β, β, y are members of T and such that the statement "β constructs M_β" is valid when relativized to T.

<u>Proof</u>. Let us note that the statement "β constructs M_β" is a sentence in ZF which says that β is an ordinal and there is a function f defined for all $\gamma < \beta$ such that for all γ, $f(\gamma) = (\bigcup_{\delta < \gamma} f(\delta))'$ and $f(\beta) = M_\beta$. This lemma is then an immediate consequence of the Löwenheim-Skolem theorem as given in Chapter II, § 8 where we take as the set S the union of M_α, $\{M_\beta\}$, $\{\beta\}$, and $\{y\}$ and consider the statements "β constructs M_β" and the Axiom of Extensionality.

By the result of Chapter II, § 6, we know that there is a unique ϵ-isomorphism φ of T onto a transitive set R. The map φ is the identity on M_α since M_α is already transitive. Also $\varphi(y) = y$ since $y \subseteq x$ and φ is the identity on the members of x. However, φ need <u>not</u> be the identity on β. Let $\varphi(\beta) = \beta'$. Then β' is an ordinal relative to R. By the absoluteness of the ordinal property, β' is actually an ordinal and clearly since $\beta' \subseteq R$ by the transitivity of R, $\bar{\bar{\beta}}' \leq \bar{\bar{R}}$ and so $\bar{\bar{\beta}}' \leq \bar{\bar{\alpha}}$. Thus "$y \in M_{\beta'}$" holds when relativized to R and by absoluteness, $y \in M_{\beta'}$ holds. This is precisely the theorem.

One should compare the above proof with the intuitive motivation. The "collapsing" of the set T by the isomorphism φ means that we have extracted from the ordinal β all the lower ordinals which played a role in the formation of y in M_β, and then discarded the other lower ordinals. The result is the much smaller ordinal β' which also contructs y.

We promised the reader in § 2 a proof which would not use the Replacement Axiom and unfortunately the Löwenheim-Skolem theorem of Chapter II, § 8 does use this axiom. However due to the very predicative nature of the statement "β constructs M_β", one can directly prove Lemma 1 or at least reduce it to the usual Löwenheim-Skolem theorem where one works within a given set. To give our alternate proof of Lemma 1, we first prove, without the Replacement Axiom,

LEMMA 2. There is a transitive set T, $M_\alpha \subseteq T$, $\{y,\beta,M_\beta\} \subseteq T$ and such that the definition of M_β holds when relativized to T.

Once Lemma 2 has been proved, the usual form of the Löwenheim-Skolem theorem as given in Chapter I, which works entirely within T and thus uses only the Separation Axiom, immediately implies Lemma 1. It is only necessary to find a T such that β, $M_\beta \in T$ and the definition of M_β holds in T, since we can then merely adjoin M_α and y to T and by the absoluteness of M_β, not alter the situation. Now the definition of M_β may not quite be capable of being given in M_β but can be in M_γ with γ only slightly larger than β. We give a short sketch which the reader should have no trouble in elaborating. To see this we first investigate the relation $Y = X'$. We claim that if $X \in M_\alpha$ for some α, then $X' \in M_{\alpha+\omega}$ and the relation $Y = X'$ is valid when relativizied to $M_{\alpha+\omega}$. The sets X_n which are used to define Y certainly occur in $M_{\alpha+\omega}$ since they are defined recursively by predicative definitions. The sets $\{\langle 1,X\rangle, \ldots, \langle n,X_n\rangle\}$ which occur in the definition of the X_n also are in $M_{\alpha+\omega}$. Therefore the definition of $n \to X_n$ can be given in $M_{\alpha+\omega}$ and hence the definition of Y can be given in $M_{\alpha+\omega}$.

Let us now recall that the set M_α is defined as $F_\alpha(\alpha)$ where F_α is the unique function defined for all $\beta \leq \alpha$ such that $F_\alpha(0) = \emptyset$ and $F_\alpha(\beta) = (\bigcup_{\gamma < \beta} F_\alpha(\gamma))'$. Let $\varphi(\alpha)$ be the function defined by $\varphi(0) = 2$ and $\varphi(\alpha) = \{\sup_{\beta<\alpha} \varphi(\beta)\} + \omega$, so that φ is strictly increasing. We show by induction that $F_\alpha \in M_{\varphi(\alpha)}$, $X_\alpha = (\bigcup_{\beta<\alpha} M_\beta) \in M_{\varphi(\alpha)}$, and that the relation $X'_\alpha = M_\alpha$ holds when relativized to $M_{\varphi(\alpha)}$. Assume this for all $\beta < \alpha$ and let $\gamma = \sup_{\beta<\alpha} \varphi(\beta)$. Then for $\beta < \alpha$, $F_\beta \in M_\gamma$. Now consider the following definition <u>relativized to</u> M_γ which therefore defines a set $S \in M_{\gamma+1}$:

"S is the union of all functions G_β, for some $\beta < \alpha$, such that G_β is defined for all $\beta' \leq \alpha$ and $G_\beta(0) = \emptyset$ and $G_\beta(\beta') = (\bigcup_{\beta''<\beta'} G_{\beta'})'$."

Because of our inductive hypothesis $S = \{\langle \beta, F_\alpha(\beta)\rangle \mid \beta < \alpha\}$.

Since $M_\alpha \in M_\gamma$ trivially, we easily get that $F_\alpha = S \cup \{\langle \alpha, M_\alpha\rangle\}$ is in $M_{\gamma+3}$. Also X_α which is the sum set of the range of S must also be in $M_{\gamma+3}$ since it can be defined predicatively from S and by the above $X'_\alpha = M_\alpha$ must hold in $M_{\gamma+3+\omega} = M_{\varphi(\alpha)}$. Thus our assertion follows and Lemma 2 is proved.

From Theorem 1 it is now trivial to prove GCH. First observe that for all infinite α, $\overline{\overline{M}}_\alpha = \overline{\overline{\alpha}}$. This is proved by induction on α, the case $\alpha = \aleph_0$ being obvious. For general α, by induction we see that if $X_\alpha = \bigcup\{M_\beta \mid \beta < \alpha\}$, then $\overline{\overline{X}}_\alpha \leq \overline{\overline{\alpha}} \cdot \overline{\overline{\alpha}} = \overline{\overline{\alpha}}$. The number of formulas used to define elements of M_α has therefore cardinality $\leq \overline{\overline{\alpha}}$ (using AC which is justified). Thus the cardinality of M_α is $\leq \overline{\overline{\alpha}}$, but since $\beta \in M_{\beta+1}$ we easily see that $\overline{\overline{M}}_\alpha \geq \overline{\overline{\alpha}}$ so $\overline{\overline{M}}_\alpha = \overline{\overline{\alpha}}$. Now if α is a cardinal, $\alpha \in M_{\alpha+1}$, and we see that if β is the next cardinal after α, $P(\alpha) \subseteq M_\beta$ so $\overline{\overline{P(\alpha)}} \leq \overline{\overline{\beta}}$. Cantor's theorem says $\overline{\overline{P(\alpha)}} > \overline{\overline{\alpha}}$ so $\overline{\overline{P(\alpha)}} = \overline{\overline{\beta}}$

which is exactly GCH. It is more usually written $2^{\aleph_\alpha} = \aleph_{\alpha+1}$ which means precisely the same.

Thus we have now proved completely Theorems 1, 2 and 3 of § 1. Although ZF is an infinite axiom scheme, for any axiom A_n we have shown how to give a proof of $A_{n,L}$ in ZF and a closer analysis will show that, under any natural enumeration of proofs P_m, the proof of $A_{n,L}$ is given by $P_{\varphi(n)}$ when $\varphi(n)$ is a suitable primitive recursive function. Thus, if we view ZF merely as a formal system we have given a proof of

$$(\text{Consis ZF}) \rightarrow (\text{Consis ZF + AC + GCH})$$

wholly within Z_1 or even a still simpler version of number theory. That is, given a contradiction in ZF + AC + GCH, we have a totally effective process for transforming it into a contradiction in ZF.

5. RELATIONS WITH GB

In his 1940 monograph [14], Gödel presented his proof for the system GB. Although GB, in our opinion, is a less intuitive system, it has the merit of being finitely axiomatizable. Indeed it can be thought of as ZF modified only by an axiomatization of the metamathematical notion of "property". He also replaced the informal definition of X', which can be made precise in many ways, one of which we have sketched, by an absolutely formal and correct definition. To do this requires an analysis of how formulas are constructed, which is precisely what is done by some of the axioms of GB. Thus, this is another reason why GB simplifies the formal presentation. Of course, the price must be paid, as he does in [14], by showing that the classes defined by GB do enable us to reconstruct any formula.

Thus, he defines a transfinite sequence of sets F_α, amongst which are to be found the M_α. In general we have $M_\alpha = F_{\varphi(\alpha)}$ where $\varphi(\alpha)$ is some explicit function of α. Each F_α will be a single set whose elements are taken from $\{F_\beta | \beta < \alpha\}$. (Recall that each M_α was a set each of whose members x was a subset of $\bigcup_{\beta<\alpha} M_\beta$.) The ordering of the F_α is somewhat arbitrary and it is not immediately clear (though essentially trivial) that all the sets M_α are obtained. Because of the predicative nature of the construction of the F_α it is quite clear that they are absolute and since the operations used to define them can be done in L, they themselves are in L. Due to the historical interest in [14] we shall now give the definition of the F_α.

<u>Definition.</u>

$$\mathfrak{F}_1(x,y) = \{x,y\}$$
$$\mathfrak{F}_2(x,y) = \{\langle r,s\rangle \mid \langle r,s\rangle \in x \ \& \ r \in s\}$$
$$\mathfrak{F}_3(x,y) = x - y$$
$$\mathfrak{F}_4(x,y) = \{\langle r,s\rangle \mid \langle r,s\rangle \in x \ \& \ s \in y\}$$
$$\mathfrak{F}_5(x,y) = \{s \mid s \in x \ \& \ \exists r(\langle r,s\rangle \in y)\}$$
$$\mathfrak{F}_6(x,y) = \{\langle r,s\rangle \mid \langle r,s\rangle \in x \ \& \ \langle s, r\rangle \in y\}$$
$$\mathfrak{F}_7(x,y) = \{\langle r,s,t\rangle \mid \langle r,s,t\rangle \in x \ \& \ \langle r,t,s\rangle \in y\}$$
$$\mathfrak{F}_8(x,y) = \{\langle r,s,t\rangle \mid \langle r,s,t\rangle \in x \ \& \ \langle t,r,s\rangle \in y\}$$

The definition of $\mathfrak{F}_i$ parallels perfectly the class axioms for GB with the exception that in $\mathfrak{F}_2$ and $\mathfrak{F}_4, \ldots, \mathfrak{F}_8$ one demands that $\mathfrak{F}_i(x,y) \subseteq x$. This is done to insure the predicative and transitive nature of the F_α, i.e, that each $F_\alpha \subseteq (\bigcup_{\beta < \alpha} F_\beta)$. The next step is to give a definite procedure for performing the operations successively that is, for each α, β, i there should be a γ such that F_γ is defined as $\mathfrak{F}_i(F_\alpha, F_\beta)$. However, there is still another operation which must be performed. To illustrate, assume we wish to obtain $F_{\alpha_1} \times F_{\alpha_2}$ in the series of the sets F_α. It is clear that by repeating the operation $\mathfrak{F}_1$ sufficiently often, we will reach an ordinal β such that if $x \in F_{\alpha_1}$, $y \in F_{\alpha_2}$ then for some $\gamma < \beta$, $F_\gamma = \langle x,y\rangle$. However, to form $F_{\alpha_1} \times F_{\alpha_2}$ we must now collect all these couples $\langle x,y\rangle$. Because of the operation $\mathfrak{F}_4$ it will not be necessary for us to collect exactly the pairs $\langle x,y\rangle$ (although we could do this). Rather for certain α, occurring arbitrarily far along in the sequence of ordinals, we define F_α simply as $\{F_\beta \mid \beta < \alpha\}$. We now make the order of the operations precise.

<u>Definition</u>. $\langle\alpha,\beta\rangle \prec \langle\alpha',\beta'\rangle$ if either $\max(\alpha,\beta) \prec \max(\alpha',\beta')$ or $\max(\alpha,\beta) = \max(\alpha',\beta')$ and either $\beta < \beta'$ or $\beta = \beta'$ and $\alpha < \alpha'$.

For $0 \leq i, j \leq 8$, put $\langle i,\alpha,\beta\rangle \prec \langle j,\alpha',\beta'\rangle$ if $\langle\alpha,\beta\rangle \prec \langle\alpha',\beta'\rangle$ or $\langle\alpha,\beta\rangle = \langle\alpha',\beta'\rangle$ and $i < j$.

It is obvious for each $\langle i,\alpha,\beta\rangle$ there is a set S of all triples which precede it under $\prec$ and S is well-ordered under $\prec$. S is a set since it is contained in $\{\langle i,\alpha',\beta'\rangle \mid \alpha',\beta' \leq \max(\alpha,\beta),\ i \leq 8\}$. Thus S can be mapped uniquely in an order presering fashion onto an ordinal. Thus, we have a formula $A(x,y)$ in ZF which defines a function $J(i,\alpha,\beta)$ taking ordinal values such that J is order preserving with respect to the ordering just introduced and if J takes the value γ and $\gamma' < \gamma$ then it also takes the value γ. Intuitively, $J(i,\alpha,\beta)$ is the length of the sequence of all triples preceding $\langle i,\alpha,\beta\rangle$. It is obvious that $J(i,\alpha,\beta) \geq \alpha$ so that J must take <u>all</u> ordinal values. This means that there are functions N, K_1, K_2 also defined by formulas in ZF such that if $J(i,\alpha,\beta) = \gamma$ then $N(\gamma) = i$, $K_1(\gamma) = \alpha$ and $K_2(\gamma) = \beta$. Since $J(0,\alpha,\beta) \geq \alpha$, it follows that $J(i,\alpha,\beta) > \alpha,\beta$ if $i \neq 0$ or equivalently $N(\gamma) \neq 0 \rightarrow K_1(\gamma) < \gamma$ and $K_2(\gamma) < \gamma$.

<u>Definition</u>. The sets F_α are defined by transfinite induction as follows:

1) $F_0 = \emptyset$

2) if $N(\alpha) = 0$ then $F_\alpha = \{F_\beta \mid \beta < \alpha\}$

3) if $N(\alpha) = i > 0$, $K_1(\alpha) = \beta$, $K_2(\alpha) = \gamma$ then $F_\alpha = \mathfrak{F}_i(F_\beta, F_\gamma)$.

We first remark that the map $\alpha \rightarrow F_\alpha$ has been defined by a formula in ZF which merely made a simple use of transfinite induction. It is also an easy exercise to show by induction that there is a simple function $\varphi(\alpha)$ such that the set $G_\alpha = \{\langle\beta,F_\beta\rangle \mid \beta \leq \alpha\}$ is in $M_{\varphi(\alpha)}$, and the functions $J(i,\alpha',\beta')$, $K_1(\alpha')$, $K_2(\alpha')$, $N(\alpha')$ are all defined correctly when we relativize to $M_{\varphi(\alpha)}$. In this way, one sees that the F_α are constructible. (One uses here the easily verifiable absoluteness of J, K_1, K_2 and N and the argument resembles the proof that $\alpha \rightarrow M_\alpha$ is definable in some M_β. If the rule for the order in which the $\mathfrak{F}_i$ are performed were not of the simple, predicative type we have given, but were highly "non-constructive" in character, it might well be false that all $F_\alpha \in L$.)

We next show that conversely each M_α is of the form F_β. If we accept the result of [14] that the sets F_α form a model for ZF, then by the absoluteness of M_α, since they can be defined in the model of the F_α, the corresponding sets are equal to M_α and here the M_α are in $\{F_\alpha\}$. However, we prefer to sketch a simple direct proof. We observe the frequently used fact that $x \in F_\alpha \to \exists_\beta\ (\beta < \alpha \ \&\ x = F_\beta)$.

LEMMA. $\forall \alpha, \beta\ \exists \gamma\ (F_\gamma = F_\alpha \times F_\beta)$.

<u>Proof</u>. Let $\delta = \sup\{J(1,\alpha',\beta') \mid \alpha',\beta' < \max(\alpha,\beta)\}$. Thus $x \in F_\alpha$, $y \in F_\beta$ implies $\{x,y\} \in \{F_{\delta'} \mid \delta' \leq \delta\}$. If $\delta_1 = \sup\{J(1,\alpha',\beta') \mid \alpha',\beta' \leq \delta\}$, then it is clear that $F_\alpha \times F_\beta \subseteq \{F_{\delta'} \mid \delta' \leq \delta_1\}$. Let $\delta_2 = J(0,\delta_1,\delta_1)$ so $\delta_2 \geq \delta_1$ and $F_\alpha \times F_\beta \subseteq F_{\delta_2}$. Let $\delta_3 = J(4,\delta_2,\beta)$ so $F_\alpha \times F_\beta \subseteq F_{\delta_3}$ and $z \in F_{\delta_3} \to z = \langle x,y\rangle$ with $y \in F_\beta$. Reversing the roles of α and β yields δ_4 such that $F_\beta \times F_\alpha \subseteq F_{\delta_4}$ and $z \in F_{\delta_4} \to z = \langle x,y\rangle$ with $y \in F_\alpha$. Let $\delta_5 = J(6,\delta_2,\delta_4)$ so $F_\alpha \times F_\beta \subseteq F_{\delta_5}$ and $z \in F_{\delta_5} \to z = \langle x,y\rangle$ with $x \in F_\alpha$. Using $\mathfrak{F}_3$ we obtain a δ_6 such that $F_{\delta_6} = F_{\delta_3} \cap F_{\delta_5} = F_\alpha \times F_\beta$.

LEMMA. $\forall \alpha_1,\ldots,\alpha_n\ \exists \beta\ (F_{\alpha_1} \times \cdots \times F_{\alpha_n} = F_B)$.

LEMMA. $\forall \alpha_1,\ldots,\alpha_n\ \exists \beta\ (F_\beta = \{\langle x_1,\ldots,x_n\rangle \mid x_i \in F_{\alpha_i}\ \&\ x_1 \in x_2\})$.

<u>Proof</u>. Use $\mathfrak{F}_2$.

LEMMA. If $F_\beta \subseteq F_{\alpha_1} \times \cdots \times F_{\alpha_n}$ and σ is any permutation of $1, \ldots, n$ then $\exists \gamma$ such that $z \in F_\gamma \leftrightarrow z = \langle x_1,\ldots,x_n\rangle$ and $\langle x_{\sigma(1)}, \ldots, x_{\sigma(n)}\rangle \in F_\beta$.

<u>Proof</u>. Use $\mathfrak{F}_6$, $\mathfrak{F}_7$ and $\mathfrak{F}_8$.

LEMMA. If $F_\beta \subseteq F_{\alpha_1} \times \cdots F_{\alpha_n}$ $\exists \gamma_1$ and γ_2 such that

1) $z \in F_{\gamma_1} \leftrightarrow z = \langle x_2,\ldots,x_n\rangle\ \&\ \exists x_1 \in F_{\alpha_1}$ such that $\langle x_1,\ldots,x_n\rangle \in F_\beta$.

2) $z \in F_{\gamma_2} \leftrightarrow z = \langle x_2,\ldots,x_n\rangle\ \&\ \forall x_1 (x_1 \in F_{\alpha_1} \to \langle x_1,\ldots,x_n\rangle \in F_\beta)$.

Proof. Use $\mathfrak{F}_3$ and $\mathfrak{F}_5$.

These lemmas now clearly imply the following theorem.

THEOREM. Let $A(x,y_1,\ldots,y_n)$ be a formula in which every bound variable t_i is restricted to a set F_{β_i}. Then for all α_i, there is a γ such that $F_\gamma = \{x \in F_{\alpha_0} | A(x,F_{\alpha_1},\ldots, F_{\alpha_n})\}$.

Let us introduce the notation "$x \in L'$" to mean $\exists\, \alpha\ (x = F_\alpha)$. We know that $L' \subseteq L$ and we shall now show $L \subseteq L'$. The above theorem implies that if $X \in L'$ and $y \in X'$ then $y \in L'$. By taking the supremum over the elements of X', it follows that for some α, $X' \subseteq \{F_\beta | \beta \leq \alpha\}$ and if $\alpha_1 = J(0,\alpha,\alpha)$ then $X' \in F_{\alpha_1}$. A simple argument of the type used in our discussion of absoluteness shows that for some α_2 the definition of X' is correct when relativized to F_{α_2}. Assume that for all $\beta < \alpha$ there is a $\varphi(\beta)$ such that the definition of M_β is correct when relativized to $F_{\varphi(\beta)}$. Let $\gamma = \sup\{\varphi(\beta) | \beta < \alpha\}$, and $\gamma_1 = J(0,\gamma,\gamma)$. In F_{γ_1} we can define $X = \bigcup_{\beta < \alpha} M_\beta$, so that $X \in L'$ and hence $M_\alpha = (\bigcup_{\beta < \alpha} M_\beta)'$ is in L'. By the above we know that M_α is correctly defined in some F_{γ_2}. Thus $L \subseteq L'$ and the two notions of constructible sets are the same. The proofs of the main theorems with the F_α are of course essentially identical to the ones we have given.

We close this discussion of Gödel's proof with the following obvious result which however may sometimes be quite useful.

THEOREM. If A is a statement involving sets of finite rank alone and A can be proved from $ZF + (V = L)$, then it can be proved from ZF alone.

Proof. It is easy to see that the notion of an integer is absolute and hence so is that of a set of finite rank. Now A is true in L since $V = L$ holds in L. The sets of finite rank in L are precisely the same as the actual sets of finite rank and hence A is true. The theorem of course cannot be extended fully to sets of countable rank since the notion "countable" is not absolute.

6. THE MINIMAL MODEL

We recall that axiom SM introduced in Chapter II, § 7 says that there is a standard model for ZF. Assuming this we prove the existence of a smallest such model.

THEOREM. ZF + SM implies the existence of a unique transitive model M such that if N is any standard model there is an ϵ-isomorphism of M into N. M is countable.

Proof. Let N be any standard transitive model. Let α_0 be the supremum of all the ordinals in N. (Since ordinal is an absolute notion we do not need to distinguish between ordinals relative to N and ordinals less than α_0.) Also if $\alpha < \alpha_0$, M_α has the same meaning when relativized to N. Let us write X_α in place of $\cup\{M_\beta | \beta < \alpha\}$. Now, since N is a model for ZF, we know X_{α_0} is a model for ZF. Let α_1 be the least ordinal such that $X_{\alpha_1} = M$ is a model for ZF. We claim M is the minimal model. For, let N be an arbitrary standard transitive model. (Assuming N is transitive is no restriction since N will always be ϵ-isomorphic to a transitive set.) As before let α_0 be the supremum of all ordinals in N that X_{α_0} is a model for ZF and so $\alpha_1 \leq \alpha_0$. Now X_{α_0} is the set of all constructible sets relative to N so $X_{\alpha_0} \subseteq N$ and thus $X_{\alpha_1} \subseteq X_{\alpha_0} \subseteq N$ and we are done. The fact that M is countable follows from the Löwenheim-Skolem theorem which guarantees the existence of countable submodels of M.

COROLLARY. The model M does not contain any set which is a standard model, i.e., SM is false in M. Also, V = L holds in M.

Proof. Since $M = \cup\{M_\beta | \beta < \alpha_1\}$, clearly V = L holds in M. If M had a standard model it would also contain a standard transitive model N since the theorem of Chapter II, § 6 holds in M. By the minimal property of M, N must contain M which is absurd.

There is another point of view which leads to the minimal model and was first expounded in [3]. Let us define a new sequence of sets

$\tilde{M}_\alpha$. Let $\tilde{M}_0 = \emptyset$ and $\tilde{M}_\alpha = (\bigcup_{\beta < \alpha} \tilde{M}_\beta)'$ where X' now denotes the following operation: For any set X, X' consists of the union of X and all sets of the following form, where all statements and definitions are taken relativized to X.

1 $\emptyset$

2 $\{x,y\}$ where $x, y \in X$.

3. the sum set of x if $x \in X$.

4. $0,1,2,\ldots,\omega$

5. the set of all y in X such that $y \subseteq x$, for a fixed x in X.

6. Let φ be single valued function in X defined by a formula relativized to X using as constants sets in X. If $u \in X$, then range of φ on u is in X'.

One observes how each of these operations corresponds to an axiom of ZF. Unlike Gödel's construction we do not have in general $X \in X'$. Indeed allowing $X \in X'$ would somehow correspond to an attempt to have the existence of a "universal" set as an axiom. For this reason Gödel's M_α never terminates. Let us call the sets in our new series $\tilde{M}_\alpha$, "strongly constructible" sets. One can now show by essentially the same arguments as for L that the axioms of ZF hold for the strongly constructible sets. Now if $\tilde{M}_\alpha$ is a model for ZF then clearly $\tilde{M}_{\alpha+1} = \tilde{M}_\alpha$ and more generally $\tilde{M}_\beta = \tilde{M}_\alpha$ for all $\beta > \alpha$. It is a simple step to show using the absoluteness of the construction that the minimal model M consists precisely of the strongly constructible sets.

THEOREM. For every element x in M there is a formula $A(y)$ in ZF such that x is the unique element in M satisfying $A_M(x)$. Thus in M every element can be "named".

Proof. Enumerate all formulas $A_n(x)$ with one free variable. Let $B_n(x)$ be the formula "$A_n(x)$ and $\exists\,\alpha$ such that $x = F_\alpha$ and for all $\beta < \alpha$, $\sim A_n(F_\beta)$". Thus $B_n(x)$ picks out at most one set x. In M all sets are constructible so that if we relativize the statements B_n to M, as we henceforth do, if there is any set x such that $A_n(x)$

there will be one such that $B_n(x)$. Now the proof of the Lowenheim-Skolem theorem shows that $M' = \{x \mid \exists n\, B_n(x)\}$ is a model for ZF which is elementarily equivalent to M. Let φ be the ϵ-isomorphism of M' onto a transitive set M''. By the minimality of M and the fact $M'' \subseteq M$ we have $M = M''$. We say $\varphi(t) = t$ for all t in M'. If not, let t satisfy $B_n(x)$. Then t satisfies B_n relativized to M' since M' is elementarily equivalent to M and hence $\varphi(t)$ satisfies B_n relativized to $M'' = M$. But there is only one t satisfying B_n in M so $\varphi(t) = t$ and $M' = M$ and the theorem is proved.

The minimal model was first investigated by Shepherdson who also used it to analyze the scope of the method of inner models. His aim was proof-theoretic rather than model-theoretic. We shall give one of his results concerning the possibility of proving the independence of CH in the next chapter.

CHAPTER IV

THE INDEPENDENCE OF THE CONTINUUM HYPOTHESIS AND THE AXIOM OF CHOICE

1. INTRODUCTION

Our main object in this chapter is to prove that CH cannot be proved from ZF (with AC included), and that AC cannot be proved from ZF. Together with the results of Chapter III this will give a complete proof of the independence of CH and AC. There are further questions which can be raised concerning the relative strength of various forms of AC as well as various questions in cardinal arithmetic and some of these will also be discussed. The general method we introduce is applicable to a wide range of independence questions and since the subject is developing rapidly we shall only be able to give an incomplete résumé of the results that are known. By ZF we shall mean the usual axioms without AC. We have already shown that $V = L \rightarrow GCH$ and $V = L \rightarrow AC$. We shall also prove that $GCH \rightarrow AC$ so that we have $V = L \rightarrow GCH \rightarrow AC$. In this chapter we show that the arrows cannot be reversed, i.e., AC is not derivable from ZF, ZF + AC does not imply GCH and ZF + GCH does not imply $V = L$.

The most natural way to give an independence proof is to exhibit a model with the required properties. This is not the only way to proceed since one can attempt to deal directly and analyze the structure of proofs. However such an approach to set-theoretic questions is unnatural since all our intuition comes from our belief in the natural, almost physical, model of the mathematical universe. Indeed with regard to ZF it is hard to conceive directly of any other models. Gödel's proof yields a new model, the class of constructible sets. For a long time this model and rather simple modifications of it were the main examples of standard models for ZF. (Of course, one also has the natural models obtained by considering sets whose rank is less than a large cardinal.) In attacking the independence of CH, the first question is to decide whether to search for standard or non-standard models. Since the negation of CH or AC may appear to be somewhat unnatural one might think it hopeless to look for standard models. However, we make a firm decision at the point to consider only standard models. Although this may seem like a very severe limitation in our approach it will turn out that this very limitation will guide us in suggesting possibilities. We begin with a negative result, due to Shepherdson [23]. We recall that in Chapter 3 we found a formula $A(x) \equiv x$ is constructible, such that the "inner model" of all x satisfying $A(x)$ was the required model. We show that this method will not apply in the present situation.

THEOREM. Let $A(x)$ be any formula in ZF. One cannot prove in ZF that the axioms of ZF and $\sim V = L$ hold when relativized to the class of all x satisfying $A(x)$. A fortiori the same applies to ZF + $\sim$ AC and ZF + AC + $\sim$ GCH.

Proof. Assume there were such an $A(x)$. Let M be the minimal model and $M' = \{x \mid x \in M \text{ and } A_M(x)\}$. Then since ZF holds for M, we would have that M' is a model for ZF in which $\sim V = L$. Since $M' \subseteq M$, we know that M' is isomorphic to M and since $V = L$ holds in M, it holds in M' which is a contradiction. One can avoid appealing to the existence of the minimal model as a "set" by considering the formula $B(x) \equiv \forall M$ (if M is a standard transitive model, $x \in M$). In effect one treats the minimal model as a class and in this way one avoids Axiom SM.

Having shown that the method of inner models fails for ZF, it is natural to see whether in an extended version of ZF one can prove the existence of an inner model for ZF + $\sim V = L$. (The above proof shows that for any "natural" extension (ZF)' of ZF, one cannot show in (ZF)' the existence of an inner model for (ZF)' + $\sim V = L$.) The simplest possibility is ZF + SM since SM asserts the existence of a standard model for ZF. Here too, we encounter an interesting limitation which remained unnoticed until after the independence problem was settled.

THEOREM. From ZF + SM or indeed from any axiom system containing ZF which is consistent with $V = L$, one cannot prove the existence of an uncountable standard model in which AC is true and CH is false, nor even one in which AC holds and which contains non-constructible real numbers.

Proof. Whereas the existence of countable models for ZF was in general regarded as a curiosity this theorem shows that they must be considered if our program is to succeed. Also, as a corollary to the Completeness Theorem one knows that a consistent system has models of arbitrarily large cardinality. In our theorem we only refer, of course, to standard models. We are implicitly using Consis ZF, of course, since otherwise all statements are provable. Since we know Consis ZF $\rightarrow$ Consis(ZF + V = L) it clearly suffices to prove that ZF + V = L implies that there are no uncountable models with the properties of the theorem. Let M be an uncountable standard transitive model, $\alpha_0 = \sup\{\alpha | \alpha \in M\}$ and assume AC holds in M. If α_0 is uncountable M contains all the countable ordinals. If α_0 is countable, let R_β be the set of elements of M of rank β. Then for some $\beta < \alpha_0$, R_β is uncountable, and by the absoluteness of the rank, R_β is definable in M. Thus M contains a set which is actually uncountable and by AC this set can be well-ordered in M, so M contains an uncountable ordinal which contradicts our hypothesis. Thus in either case M contains all countable ordinals. From V = L, we know that every real number is constructible from a countable ordinal, so we have shown that in M, every real number is constructible. This proves the second assertion of our theorem and the following lemma gives the first assertion.

LEMMA. ZF + (every real number is constructible) $\rightarrow$ CH.

Proof. Our hypothesis implies that the set of all real numbers, C, is in L. In L, C has the cardinality of $\aleph_1$. This means there is a map from the countable ordinals relative to L onto C, and since a countable ordinal in L is certainly countable, there is a map from $\aleph_1$ onto C.

2. INTUITIVE MOTIVATION

To fulfill our program and show the existence of a standard model in which AC is false, or in which AC and $\sim$ CH hold, we certainly need to assume Axiom SM. Therefore, until further notice we assume SM. Also, the word model will henceforth be reserved only for standard transitive models. Our last theorem shows that we must consider countable models. Let M thus denote for the remainder of this chapter a fixed countable model for ZF, which we take for definiteness as the minimal model. Then $V = L$ holds in M. Our first goal is to see what other standard models we have any hope of showing exist. Let $\alpha_0 = \sup\{\alpha | \alpha \in M\}$. We use the notation M_α as in Chapter III to denote the constructible sets constructed at the α-th stage of the construction. We have then $M = \cup\{M_\beta | \beta < \alpha_0\}$. The following two theorems assume the consistency of ZF + SM since they use the minimal model.

THEOREM. It is consistent to assume that for no $\alpha > \alpha_0$ is $\cup\{M_\beta | \beta < \alpha\}$ a model for ZF.

Proof. If for all $\alpha > \alpha_0$, $\cup\{M_\beta | \beta < \alpha\}$ is not a model for ZF we are done. If not, let α_1 denote the least ordinal for which it is a model and put $N = \cup\{M_\beta | \beta < \alpha_1\}$. Then clearly in N it is true that $\alpha > \alpha_0 \rightarrow M_\alpha$ is not a model for ZF and the theorem is proved.

COROLLARY. It is consistent to assume that for any model N, $\alpha_0 = \sup\{\alpha | \alpha \in N\}$.

Proof. If N' denotes the constructible sets of N, then $\sup\{\alpha | \alpha \in N'\} = \sup\{\alpha | \alpha \in N\}$ and the theorem implies that it is consistent to assume $\sup\{\alpha | \alpha \in N'\} = \alpha_0$.

Thus, whatever new model N we construct, we must make do with the ordinals in M and our only hope is to construct N by "broadening" M by introducting new sets of rank α where $\alpha < \alpha_0$. Since the sets of finite rank are absolute the first possibility is to find a model N such that for some $a \subseteq \omega$, $a \in N$, and $\sim a \in M$. If N contains a, it must also contain all sets constructible from it. That is, put $M_0(a) = \omega \cup \{a\}$ so that $M_0(a)$ is transitive and put $M_\alpha(a) = (\bigcup_{\beta<\alpha} M_\beta(\alpha))'$ for $\alpha > 0$. Clearly then $\alpha \in M$ implies $M_\alpha(a) \in N$. Since these are the only sets which we can see must belong to N, it is not unreasonable to seek N to be of the form $\cup\{M_\beta(a) | \beta < \alpha_0\}$ for some $a \subseteq \omega$. If $\sim a \in M$, then we will certainly have that a is not constructible in N. This will give a model in which V = L fails. Since having ~ AC or AC + ~ CH involves a technically more complicated construction, we first concentrate on finding such an a. Observe, that if we take an arbitrary $a \subseteq \omega$ not in M, then in general $N = \cup\{M_\beta(a)\ \beta < \alpha_0\}$ will not be a model for ZF. For, the countable ordinal α_0, which is not in M, corresponds to a well-ordering of ω, and thus to a subset of $\omega \times \omega$, and hence to a subset of ω itself. If a is taken as that subset, then any model for ZF containing a must contain α_0, since it is a theorem of ZF that to every well-ordering there corresponds a unique ordinal. However, clearly $x \in N \rightarrow \text{rank } x < \alpha_0$ so we cannot have $\alpha_0 \in N$. Thus a must have certain special properties if N is to be a model. Rather than describe a directly, it is better to examine the various properties of a and determine which are desirable and which are not. The chief point is that we do not wish a to contain "special" information about M, which can only be seen from the outside, such as the countability of α_0, and will imply that any model containing a must contain more ordinals than those in M. The a which we construct will be referred to as a "generic" set relative to M. The idea is that all the properties of a must be "forced" to hold merely on the basis that a behaves like a "generic" set in M. This concept of deciding when a statement about a is "forced" to hold is the key point of the construction. Clearly there are some properties of a which no reasonable procedure could interpret as being true or false for a "generic" set a. These are statements of the form $n \in a$ where n is a particular integer. A finite set P of statements of the form $n_k \in a$ or $\sim n_k \in a$ which is self-consistent will be known as a <u>forcing condition</u>. Given P we will then ask, if under some procedure to be given, it is reasonable to expect that P "forces" a statement A about a to hold, or "forces" ~ A to hold, or whether the conditions in P do not "force" A one way or the other. Although "forcing" will be related to the notion of "implication" it will differ from it in that given that P forces A it will not be true that any a that satisfies P also satisfies A. What will be true is that any "generic" a satisfying P will also satisfy A.

After we have made the definition of forcing precise and proven some of the elementary properties about it, we shall then show that it is possible to find an infinite sequence of forcing conditions P_n, such that $P_n \subseteq P_{n+1}$, i.e., so that the chain of P_n form a consistent family of conditions, and such that for every property A, some P_n either forces A or forces $\sim A$. Such a sequence will be called a complete sequence, and since it decides every property about a it determines a itself. Using this a we can prove that the set N is a model. It will also follow that $\sim a \in M$. Furthermore in N, a statement A will be true precisely if it is forced by some P_n in the complete sequence. Any set a which arises from this process will be called a "generic" set.

Although we have not yet given the definition of forcing, it is clear that there are some properties of it that we would like to hold. First, it should be consistent, i.e., we should not have P forces A and P forces $\sim A$. Secondly, if P forces A and $Q \supseteq P$ then Q forces A. Thirdly, for every P and A there is a Q such that $Q \supseteq P$ and either Q forces A or Q forces $\sim A$. Clearly these properties correspond to the usual properties of implication. Bearing these requirements in mind will help to understand the choice of the definition of forcing.

As further motivation, let us recall the basic idea of constructing models as embodied in the Completeness Theorem. When we try to construct a model for a collection of sentences, each time we encounter a statement of the form $\exists x\, B(x)$ we must invent a symbol $\bar{x}$ and adjoin the statement $B(\bar{x})$. In the present case we are starting with a single symbol for the set a and wish, in some sense, to give the least possible information about it. This would lead to the intuition that when faced with $\exists x\, B(x)$, we should choose to have it false, unless we have already invented a symbol $\bar{x}$ for which we have strong reason to insist that $B(\bar{x})$ be true. Of course, if $B(x)$ is of the form $\forall y\, C(x,y)$ then even if we take $\exists x\, \forall y\, C(x,y)$ as false, then for particular x, say $x = 1$, we will then have to invent a $\bar{y}$ such that $\sim C(1,\bar{y})$. These ideas of course cannot be made too precise since they are too general as they apply to arbitrary axiom systems. We must add to them the essential feature of ZF, which is the natural hierarchy of sets induced by their rank. Using this hierarchy we can, by a suitable transfinite induction, define the notion of "forcing" referred to above. The crucial idea will be the preferential treatment of the universal quantifier over the existential quantifier.

3. THE FORCING CONCEPT

Our goal is to find $a \subseteq \omega$ such that $\sim a \in M$ and if $N = \cup\{M_\beta(a) \mid \beta \in M\}$, then N is a model for ZF. The presentation we give here has some inessential variations from that of [4]. First, in [4] we used the F_α construction rather than M_α because it was only the former which was presented in the literature [14] with complete details. The M_α presentation is probably superior in most regards. Second, we incorporate several simplifications which were pointed out by various people.

As mentioned, we shall examine all possible statements about N and decide if we wish them to be true or false. This requires giving names to the elements of N before we have actually chosen a and thus before we have N explicitly. In [4], we used F_α as formal symbols which were to be interpreted as $F_\alpha(a)$ after a was chosen, and considered sentences involving these formal symbols. Thus the elements of N were given ordinal "labels" but of course, it happened that for some α, β, F_α and F_β were eventually identified. The way we assign "labels" is of no importance as long as we have not neglected any set in N. The situation is analogous to the construction of the extension of a field k formed by adjoining the root α of an irreducible equation $f(x) = 0$. The elements of the extension field are all of the form $p(\alpha)$ where p is a polynomial and α is taken as a formal symbol, but we identify $p(\alpha)$ and $q(\alpha)$ if $p(x) - q(x)$ is divisible by $f(x)$. Now, the sets in $M_\alpha(a)$ which are not in $M_\beta(a)$, $\beta < \alpha$, are all defined by formulas $\{x \in X_\alpha \mid A(x)\}$ where $X_\alpha = \bigcup_{\beta<\alpha} M_\beta(a)$ and A is a formula where all bound variables are restricted to X_α and which may involve as constants particular elements of X_α. What we need is therefore some enumeration of all such possible formulas. The F_α of [4] are merely one particular way of doing this.

<u>Definition</u>. A "labeling" is a mapping defined in ZF, which assigns to each ordinal $0 < \alpha < \alpha_0$, a set S_α, the "label space", and functions φ_α defined in S_α such that the sets S_α are disjoint and if $c \in S_\alpha$, $\varphi_\alpha(c)$ is a formula $A(x)$ which has all its bound variables restricted to X_α and which may have elements of S_β with $\beta < \alpha$ appearing as constants. The function φ_α must put S_α into one-one correspondence with the set of all such formulas. The set S_0 is defined as the set $\omega \cup \{a\}$ where a is a formal symbol. We write $S = \bigcup_\alpha S_\alpha$.

It is clear that each $c \in S$ is in a unique S_α. When we are done and finally choose the particular set $a \subseteq \omega$, we shall be able to define for each $c \in S_\alpha$ a set $\bar{c}$ as follows. For c in S_0, $\bar{c}$ is obviously defined. For c in S_α, $\alpha > 0$, let $\varphi_\alpha(c) = A(x, c_1, \ldots, c_m)$, $c_i \in S_{\beta_i}$, $\beta_i < \alpha$ where $A(x, t_1, \ldots, t_m)$ is a formula in ZF with all bound variables restricted to X_α. To indicate that a bound variable x is restricted to X_α we shall henceforth write $\forall_\alpha x$ or $\exists_\alpha x$. If we define X_α inductively as $\{\bar{c} \mid c \in S_\beta \;\&\; \beta < \alpha\}$, then $\bar{c} = \{x \in X_\alpha \mid A(x, \bar{c}_1, \ldots, \bar{c}_m)\}$. The very meticulous reader may object that in ZF one does not have symbols for "formulas" and so we cannot define φ_α. We mean, of course, that we assume a natural enumeration of all formulas $A(x, t_1, \ldots, t_m)$ with more than one variable and the function φ_α actually consists of two functions, φ_α^1 which gives the number of the formula $A(x, t_1, \ldots, t_m)$ and φ_α^2 which is defined for all integers i and such that $\varphi_\alpha^2(i) = c_i$ if $i \leq m$, $\varphi_\alpha^2(i) = 0$ if $i > m$, so that φ_α^2 takes its values in $\{0\} \cup \{\bigcup_{\beta<\alpha} S_\beta\}$.

Clearly one can define the "label" spaces S_α by transfinite induction in many ways. Perhaps the simplest way is to define S_α as the set of all formulas $A(x)$ themselves. That is we define S_α by induction as the set of all formulas $(A(x, c_1, \ldots, c_m))$ where $c_i \in \bigcup_{\beta<\alpha} S_\alpha$ and such that all quantifiers are of the form $\forall_\alpha$ or $\exists_\alpha$. This means that an element of S_α is a formula some of whose terms are themselves formulas, etc., hence has a finite nested sequence of formulas in it and in which every quantifier has an ordinal index. This is the approach taken in [6], [7]. One slight difficulty is that in ZF we have no "formal symbols" since everything is a set. The symbols we need are however finite in number, i.e., $\&$, $\sim$, $\vee$, $\rightarrow$, $\leftrightarrow$, $=$, $\in$, $\forall$, $\exists$, x, $($, $)$ so that we can assign distinct integers to represent them. Then the elements of the label space S consist of finite sequences of these integers or ordinals with the understanding that $\forall$ and $\exists$ are always followed by an ordinal so that $\forall, \alpha$ stands for $\forall_\alpha$ and x is always followed by an integer n that x, n stands for x_n. Of course only certain of these sequences correspond to well-formed formulas and elements of S but the rules for deciding this are simple and easily expressible in ZF.

We are now ready to examine the sentences about N.

__Definition__. A __limited statement__ is a statement in ZF in which every quantifier is of the form $\forall_\alpha$ or $\exists_\alpha$ for some ordinal $\alpha < \alpha_0$ and which may involve elements of S as constants.

__Definition__. An __unlimited statement__ is a statement in ZF which may involve elements of S as constants.

When N is finally constructed, the limited and unlimited statements become actual statements about N when we replace each c in S by $\bar{c}$ and $\forall_\alpha x$ and $\exists_\alpha x$ are interpreted as $\forall x$ and $\exists x$ but x are restricted to lie in X_α. The limited statements are obviously of a simpler type since they speak only about the elements of X_α for some α. They have a natural hierarchy as follows:

__Definition__. If A is a limited statement, let rank A = (α, i, r) where

1. α is the least ordinal such that if $\forall_\beta$ or $\exists_\beta$ occur in A, then $\beta \leq \alpha$ and if $c \in S_\beta$ occurs in A, then $\beta < \alpha$.
2. r is the number of symbols in A.
3. $i = 0$ if α is a successor ordinal, say $\alpha = \beta + 1$, and $\exists_\alpha$ and $\forall_\alpha$ do not occur in A, and no term of the form $c \in (\cdot)$, $c = (\cdot)$, or $(\cdot) = c$, occurs in A where $c \in S_\beta$. Otherwise $i = 1$.

If rank A = (α, i, r), A speaks only about the elements in X_α. The index r is clearly a measure of the complexity of A. The index i is necessary because when we encounter a term of the form $c_1 \in c_2$ where $c_1 \in S_\beta$ and $c_2 \in S_\gamma$ for $\gamma < \beta$, we will encounter some technical difficulties in replacing $c_1 \in c_2$ by an "equivalent" statement of a simpler type. This will appear in our definition of forcing.

__Definition__. $(\alpha_1, i_1, r_1) < (\alpha_2, i_2, r_2)$ if $\alpha_1 < \alpha_2$ or $\alpha_1 = \alpha_2$ and $i_1 < i_2$ or $\alpha_1 = \alpha_2$, $i_1 = i_2$ and $r_1 < r_2$.

__Definition__. A __forcing condition__ P is a finite set of limited statements of the form $n \in a$ or $\sim n \in a$ where $n \in \omega$ and n and a are regarded as belonging to S_0, and such that for given n, not both $n \in a$ and $\sim n \in a$ are in P.

We now wish to define "P forces A" and we consider first the case where A is a limited statement. The definition proceeds by induction on rank A. The essential step consists in reducing A to statements of lower rank. If A is of the form $\exists_\alpha x\, B(x)$ we shall say that P forces A quite simply if for some $c \in S_\beta$, $\beta < \alpha$, P forces $B(c)$. This means that in accordance with the idea that a is "generic", a set exists only if there is a fixed c such that for all "generic" a satisfying P, $B(\bar{c})$ will be true. This is a uniformity statement on a in that the set which satisfies $B(x)$ can be given a fixed definition in terms of a. Now, if A is of the form $\forall_\alpha x\, B(x)$ in accordance with our general ideas we might be tempted to say that P forces A unless for some $c \in S_\beta$, $\beta < \alpha$, P forces $\sim B(c)$. However, there are other considerations which might make us reluctant to do so. Namely, suppose that for some Q which contains P we have that Q forces $\sim B(c)$ with $c \in S_\beta$, $\beta < \alpha$. Saying then that P forces A will then mean that we will never be able to have a satisfy the conditions of Q, which would seem to violate the "generic" character of a in that a would satisfy some hidden conditions not mentioned in P. It may even occur that there are so many $Q \supseteq P$ such that for some $c \in S_\beta$, $\beta < \alpha$, Q forces $\sim B(c)$, that it would make it absurd to say that P forces A. Thus, we shall say P forces $\forall_\alpha x\, B(x)$ if for all $Q \supseteq P$, $c \in S_\beta$, $\beta < \alpha$, Q does <u>not</u> force $\sim B(c)$.

Let us look at how this works in some examples. Consider the statement "a is infinite". This says $\forall_1 x\ \exists_1 y\ (y > x \ \&\ y \in a)$. (Technically $>$ is not an admissible symbol in our formulas, but we ignore this.) For any particular n_0, no P will ever force n_0 to be an upper bound for the integers in a. Thus, <u>every</u> P forces "a is infinite". A similar argument shows that every P forces a to have infinitely many primes. Also, it is quite plausible that every P forces a to be non-constructible since for any given constructible element x one should be able to force $\sim x = a$ by a single condition of the form $n \in a$ or $\sim n \in a$. This will eventually be proved rigorously.

In the original version of forcing [2], all limited statements were thought of as being brought into prenex form. In this form the elimination of quantifiers is the only essential step since the other rules are more or less obvious. However, if one thinks of $\forall$ as being $\sim\exists\sim$, then the special treatment of $\forall$ implies a special treatment of $\sim$. In general, we say P forces $\sim A$ if for all $Q \supseteq P$, Q does not force A. This improvement, due to D. Scott, allows us to handle statements without bringing them into prenex form as well as to effect other simplifications. The models N one finally obtains are of course the same under any of these variations.

Definition. We define P forces A, for A a limited statement, by induction on rank A as follows:

1. P forces $\exists_\alpha x\, B(x)$ if for some $c \in S_\beta$, $\beta < \alpha$, P forces $B(c)$.
2. P forces $\forall_\alpha x\, B(x)$ if for all $Q \supseteq P$ and $c \in S_\beta$, $\beta < \alpha$, Q does not force $\sim B(c)$.
3. P forces $\sim B$ if for all $Q \supseteq P$, Q does not force B.
4. P forces B & C if P forces B and P forces C.
5. P forces $B \vee C$ if either P forces B or P forces C.
6. P forces $A \rightarrow B$ if either P forces B or P forces $\sim A$.
7. P forces $A \leftrightarrow B$ if P forces $A \rightarrow B$ and P forces $B \rightarrow A$.
8. P forces $c_1 = c_2$, where $c_1 \in S_\alpha$, $c_2 \in S_\beta$, $\gamma = \max(\alpha,\beta)$ if either $\gamma = 0$ and $c_1 = c_2$ as elements of S_0 or $\gamma > 0$ and P forces $\forall_\gamma\, x\, (x \in c_1 \leftrightarrow x \in c_2)$.
9. P forces $c_1 \in c_2$ where $c_1 \in S_\alpha$, $c_2 \in S_\beta$, $\alpha < \beta$ if P forces $A(c_1)$ where $A(x) = \varphi_\beta(c_2)$ (i.e., $A(x)$ is the formula defining c_2).
10. P forces $c_1 \in c_2$ where $c_1 \in S_\alpha$, $c_2 \in S_\beta$, $\alpha \geq \beta$ and not $\alpha = \beta = 0$, if for some $c_3 \in S_\gamma$, $\gamma < \beta$ if $\beta > 0$, $\gamma = 0$ if $\beta = 0$, P forces $\forall_\alpha x\, (x \in c_1 \leftrightarrow x \in c_3)$ & $(c_3 \in c_2)$.
11. P forces $c_1 \in c_2$, where $c_1, c_2 \in S_0$ if $c_1, c_2 \in \omega$ and $c_1 \in c_2$ (as elements of S_0) or $c_2 = a$ and the statement $c_1 \in a$ is in P.

In case 10, observe that if $c_1 \in S_\alpha$, $c_2 \in S_\beta$, $\alpha \geq \beta$, then if $A \equiv c_1 \in c_2$ rank $A = (\alpha+1,1,3)$. The rank of $\forall_\alpha x\, (x \in c_1 \leftrightarrow x \in c_3)$ & $(c_3 \in c_2)$ is $(\alpha+1,0,17)$ so that we have achieved a reduction in rank. Similarly in case 8 we see that the index i is reduced from 1 to 0. In the other cases if the rank of the statement is (α,i,r) it is clear that the i index is never increased and there is always either a reduction of α or r effected by the definition.

Definition. We define P forces A where A is an unlimited statement by induction on the number of symbols in A as follows:

1. P forces $\exists x\, B(x)$ if for some $c \in S$, P forces $B(c)$.
2. P forces $\forall x\, B(x)$ if for all $c \in S$, $Q \supseteq P$, Q does not force $\sim B(c)$.
3. P forces $\sim B$ if for all $Q \supseteq P$, Q does not force B.
4. P forces B & C if P forces B and P forces C.
5. P forces $B \vee C$ if P forces B or P forces C.
6. P forces $B \rightarrow C$ if P forces C or P forces $\sim B$.
7. P forces $B \leftrightarrow C$ if P forces $B \rightarrow C$ and P forces $C \rightarrow B$.
8. P forces $c_1 \in c_2$ or $c_1 = c_2$ if it forces them as limited statements.

We repeat that our definitions do not imply that for P and A we must have either P forces A or P forces $\sim A$. Also forcing does not obey some simple rules of the propositional calculus. Thus, P may force $\sim \sim A$ and yet not force A.

4. THE MAIN LEMMAS

In the following A denotes either a limited or unlimited statement.

LEMMA 1. For all P and A, we do not have both P forces A and P forces $\sim A$.

Proof. This is immediate from the definition since if P forces $\sim A$, $P \supseteq P$ so P cannot force A.

LEMMA 2. If P forces A and $Q \supseteq P$ then Q forces A.

Proof. We prove this first for limited A by induction on rank A $= (\alpha, i, r)$. Cases 4 through 11 in the definition of forcing require no discussion since P forces A is reduced to P forces B, for some B with rank B less than rank A. If P forces $\exists_\alpha x\, B(x)$ then P forces $B(c)$, $c \in S_\beta$, $\beta < \alpha$, so by induction Q forces $B(c)$ so Q forces

$\exists_\alpha x\ B(x)$. If P forces $\forall_\alpha x\ B(x)$, and $Q \supseteq P$, then if $R \supseteq Q$, $R \supseteq P$ also so R does not force $\sim B(c)$ for any $c \in S_\beta$, $\beta < \alpha$, so Q forces $\forall_\alpha x\ B(x)$. If P forces $\sim B$, $Q \supseteq P$ and $R \supseteq Q$, $R \supseteq P$, also, so R does not force B, so Q forces $\sim B$. Now, if A is unlimited the same arguments apply to Cases 1 through 7. Case 8 is handled by referring to the proof for limited statements.

LEMMA 3. For all P and A there is a $Q \supseteq P$ such that either Q forces A or Q forces $\sim A$.

<u>Proof</u>. This is the surprising fact about forcing, namely that every statement in N, which is thus a statement about a, is decidable in some sense by finitely many statements of the form $n \in a$ or $\sim n \in a$. The proof however is trivial. If P does not force $\sim A$ by definition it must be because for some $Q \supseteq P$, Q forces A.

<u>Definition</u>. A sequence $\{P_n\}$ of forcing conditions is a <u>complete sequence</u> if $P_n \subseteq P_{n+1}$ for all n, and for every A, limited or unlimited, $\exists n$ such that either P_n forces A or P_n forces $\sim A$. The sequence $\{P_n\}$ (i.e., $\{\langle n, P_n\rangle\}$) is <u>not</u> assumed to be in M.

LEMMA 4. A complete sequence exists.

<u>Proof</u>. Here for the first and only time we use the countability of M. Since M is countable we can enumerate all statements A_n. Define P_n by induction as any forcing condition $Q \supseteq P_{n-1}$ such that either Q forces A_n or Q forces $\sim A_n$. (Since the set of all P is countable one is not using AC here.)

If $\{P_n\}$ is complete then in particular for every k either some P_n forces $k \in a$ or $\sim k \in a$. Let $\bar{a} = \{k \mid \exists n\ (P_n \text{ forces } k \in a)\}$. Then $\bar{a} \subseteq \omega$ and as we remarked we can now define a map $c \to \bar{c}$ defined for all c in S which sends a into $\bar{a}$ and we can then define N as $\cup\{M_\beta(\bar{a}) \mid \beta < \alpha_0\}$. We can now connect the notion of forcing with that of truth in N.

LEMMA 5. A is true in N if and only for some n, P_n forces A.

<u>Proof</u>. Assume first that A is limited. We proceed by induction on the rank as in the definition of forcing. If A is of the form

$\exists_\alpha x\, B(x)$ and P_n forces A then P_n forces $B(c)$, $c \in S_\beta$, $\beta < \alpha$ and so by induction $B(\bar{c})$ is true in N and hence so is A. Conversely, if A is true in N then for some $c \in S_\beta$, $\beta < \alpha$, $B(\bar{c})$ holds so by induction some P_n forces $B(c)$ and so P_n forces A. If $A \equiv \forall_\alpha x\, B(x)$ and P_n forces A, $c \in S_\beta$, $\beta > \alpha$, then some P_m, $m > n$ must force $B(c)$ since no P_m can force $\sim B(c)$. By induction this means $B(\bar{c})$ is true in N, so A is true in N. Conversely, assume A is true in N. If some P_n forces $\exists_\alpha x \sim B(x)$ then for some $c \in S_\beta$, $\beta > \alpha$ P_n forces $\sim B(c)$ so $\sim B(\bar{c})$ holds in N and A is false in N. Thus, some P_n must force $\sim \exists_\alpha x \sim B(x)$, which means that for $Q \supseteq P_n$, Q does not force $\exists_\alpha x \sim B(x)$ and so for all $c \in S_\beta$, $\beta < \alpha$, Q does not force $\sim B(c)$. This says precisely that P_n forces $\forall_\alpha x\, B(x)$. If P_n forces $\sim A$ and A were true in N, by induction some P_m, $m > n$ must force A. But P_m forces $\sim A$ which contradicts Lemma 1. The other cases are trivial. If A is unlimited the same arguments go over unchanged except that the induction proceeds on the number of symbols.

5. DEFINABILITY OF FORCING

As explained in § 3, we can assign distinct integers to the finitely many symbols of our language and then every limited statement corresponds to a finite sequence of integers and ordinals with the understanding that every time $\forall$ or $\exists$ appear they are followed by an ordinal and each appearance of x is followed by an integer. These sequences must be subject to the laws of formation of well-formed formulas. For the moment, we allow arbitrary ordinals α to appear and do not demand $\alpha < \alpha_0$. The limited statements are thus objects inside ZF and there is no ambiguity in referring to them. The definition of forcing we gave was totally within the language of ZF and proceeded by a simple transfinite induction. More specifically, for each rank (α,i,r) there is a set $T_{\alpha,i,r}$ consisting of limited formulas of rank less than (α,i,r) and the definition proceeded by induction. We nowhere used the restriction that $\alpha < \alpha_0$. Thus we have a relation $F(P,A)$ expressible in ZF which means "P forces A". (We assume that each P is coded into ZF in any manner, e.g., as an ordered pair of disjoint finite sets of integers.)

LEMMA. $F(P,A)$ is an absolute relation.

We omit the proof of this lemma since it is both tedious and obvious. The definition proceeded by an induction which only required looking at "earlier" objects and this is clearly absolute. The proof is essentially the same as the proof that constructibility is absolute.

When we turn to unlimited statements, if we look at the definition of forcing without the assumption that all ordinals are less than α_0, the situation is quite different. It is then impossible to define forcing in ZF by induction because the informal induction proceeded by induction on the number of symbols, but since we allow constants $\in S_\alpha$ for arbitrary α, there is no set consisting of the statements with r symbols. Of course if we assume all $\alpha < \alpha_0$ then it can be expressed in ZF, but α_0 must appear in the defining relation. However, in perfect analogy with the result of Chapter II, § 6 we can handle a single formula.

LEMMA. Let $A(x_1,\ldots,x_n)$ be a an (unlimited) formula in ZF, involving no constants, having n free variables. There is a relation $F(P,c_1,\ldots,c_n)$ expressible in ZF which is absolute and which when relativized to any model for ZF says that P forces $A(c_1,\ldots,c_n)$.

Proof. The lemma is proved by induction on the number of symbols in A. If A has no quantifiers the result has already been proved since A is then a limited statement. If A is a propositional function of simpler statements the induction is obvious. If A is of the form $\forall y\ B(y,x_1,\ldots,x_m)$, by induction there is a statement $G(P,c_0,c_1,\ldots,c_n)$ which means that P forces $B(c_0,\ldots,c_n)$. To say that P forces $A(c_1,\ldots,c_n)$ is then simply written as $\forall Q,\ c_0(Q \supseteq P \rightarrow \sim G(Q,c_0,\ldots,c_n))$. If A is $\exists y\ B(y,x_1,\ldots,x_n)$, we merely write $F \equiv \exists c_0\ (G(P,c_0,\ldots,c_n))$.

The result can be easily generalized to handle all unlimited statements which contain less than r quantifiers, for fixed r. We have no need for this generalization.

6. THE MODEL N

Assume that a complete sequence $\{P_n\}$ has been chosen, and let N be the resulting set $\cup\{M_\beta(\bar{a}) \mid \beta < \alpha_0\}$. We now prove that N is a model for ZF. The intuition why this is so is difficult to explain. Roughly speaking, the crucial axioms of ZF (Replacement and Power Set) assert the existence of certain "big" sets. Since we have started with all ordinals less than α_0, we have enough ordinals in N unless for some reason new relations between these ordinals which did not exist in M, now appear in N. The definition of forcing was designed to prevent this. That is, no information can be extracted from the set $\bar{a}$ which was not already present in M.

The proof parallels closely the proof that the axioms of ZF hold in L. Indeed, it is possible to give a unified treatment of both cases. The verification of all the axioms with the exception of Replacement and Power Set is trivial and is left to the reader. We turn to the Power Set Axiom. The original proof in [4] paralleled the second proof given in Chapter III, since that proof gives more information and one of our primary interests was CH. However, R. Solovay pointed out that Gödel's first proof carries over quite simply to our situation and we present this proof.

Let $c_0 \in S_\alpha$, $\alpha < \alpha_0$ be fixed. We shall show that there is an element of N which is the power set of $\bar{c}_0$ relative to N. In the corresponding proof for L, we first examined the true power set and then cut down to the constructible power set. Since $\bar{c}_0$ is not necessarily in M we cannot do this as we wish to keep our argument entirely within M so that we can use the fact that M is a model for ZF. Although $\bar{a}$ is not in M, the crucial information about $\bar{a}$ is contained in the forcing conditions P, and the set of all P is a set in M. Thus we can reason about the P in M rather than the sets $\bar{c}_0$ or $\bar{a}$. In the following we return to the restriction that only $\alpha < \alpha_0$ and S_α with $\alpha < \alpha_0$ are being considered. Recall that $c_0 \in S_\alpha$ is fixed.

<u>Definition</u>. For each c in S, let $R(c) = \{P \mid P \text{ forces } c \subseteq c_0\}$, $T(c) = \{\langle P, c'\rangle \mid P \text{ forces } c' \in c \text{ and } c' \in S_\beta \text{ with } \beta < \alpha\}$, $U(c) = \langle R(c), T(c)\rangle$.

Observe that $R(c)$, $T(c)$, $U(c)$ are all sets in M, and by the results of the last section these functions are all definable by formulas in ZF <u>which are relativized to</u> M. The important point is that these definitions do not use the particular complete sequence $\{P_n\}$ which was chosen. $T(c)$ is a set because of the important restriction $\beta < \alpha$.

LEMMA. If $U(c_1) = U(c_2)$ and $\bar{c}_1 \subseteq \bar{c}_0$ then $\bar{c}_1 = \bar{c}_2$.

Proof. Since $\bar{c}_1 \subseteq \bar{c}_0$ some P_n in the complete sequence must force $c_1 \subseteq c_0$. But since $R(c_1) = R(c_2)$, P_n also forces $c_2 \subseteq c_0$, so $\bar{c}_2 \subseteq \bar{c}_0$. If $\bar{c}_1 \neq \bar{c}_2$, then for some $c_3 \in S_\beta$, $\beta < \alpha$ either $\bar{c}_3 \in \bar{c}_1$ & $\sim \bar{c}_3 \in \bar{c}_2$ or $\sim \bar{c}_3 \in \bar{c}_1$ & $\bar{c}_3 \in \bar{c}_2$. Assuming the first without loss of generality, there must be some P_n which forces $c_3 \in c_1$. Since $T(c_1) = T(c_2)$, P_n also forces $c_3 \in c_2$ so $\bar{c}_3 \in \bar{c}_2$ which is a contradiction. Thus $\bar{c}_1 = \bar{c}_2$.

Now, S is not a set in M. (It is, of course, a class.) However the range of $R(c)$ for all $c \in S$ is contained in the power set of all P, relative to M, since R is definable in M. Similarly the range of T is contained in the power set, relative to M, of the direct product of the set of all P and $\bigcup_{\beta < \alpha} S_\beta$. Thus the range of $U(c)$ is a <u>set</u> in M, $\bar{U}$. For each $u \in \bar{U}$, let $f(u)$ be the least β such that for some c in S_β, $U(c) = u$ and $f(u) = 0$ if no such β exists. Let $\beta_0 = \sup\{f(u) | u \in \bar{U}\}$, so $\beta_0 \in M$ since we have stayed entirely within the model M.

LEMMA. If $\bar{c} \subseteq \bar{c}_0$ then for some $c_1 \in S_\beta$ with $\beta < \beta_0$, $\bar{c}_1 = \bar{c}$.

Proof. By the definition of β_0 there is a $c_1 \in S_\beta$, $\beta < \beta_0$ such that $U(c) = U(c_1)$. Our previous lemma now implies $\bar{c}_1 = \bar{c}$.

THEOREM. The Power Set Axiom holds in N.

Proof. Since all the subsets in N of $\bar{c}$ are members of $X_{\beta_0} = \cup\{M_\beta(\bar{a}) | \beta < \beta_0\}$ the power set of $\bar{c}$ in N is $\{x \in X_{\beta_0} | x \subseteq \bar{c}\}$ and hence being defined by a formula relativized to X_{β_0}, is a member of $M_{\beta_0}(\bar{a})$.

We now turn to the Replacement Axiom. Again the argument parallels the corresponding argument for L. Let $A(x,y)$ be a formula which defines $y = \varphi(x)$ as a univalent function in N. Let c_0 be fixed, $c_0 \in S_\alpha$. By the previous section we can define in M the function $g(P,c)$ = least β such that for some $c' \in S_\beta$, P forces $\varphi(c) = c'$, and $g = 0$ if no such β exists. Let $\beta_0 = \sup\{g(P,c) | \text{all } P \text{ \& all } c \in S_\beta, \beta < \alpha\}$. We thus see that the range of φ on $\bar{c}_0$ is contained in $M_{\beta_0}(\bar{c})$. To obtain the range itself we need the following

THEOREM. Let $A(x_1,\ldots,x_n)$ be a formula in ZF. For each β there exists $c \in S$ such that, independently of the complete sequence $\{P_n\}$, $\bar{c} \supseteq M_\beta(\bar{a})$ and for all $\bar{x}_i \in \bar{c}$, $A_N(\bar{x}_1,\ldots,\bar{x}_n) \leftrightarrow A_{\bar{c}}(\bar{x}_1,\ldots,\bar{x}_n)$.

Proof. Recall that $A_{\bar{c}}$ means A relativized to $\bar{c}$. Assume A is of the form $Q_1y_1\ldots Q_my_m\, B(x_1,\ldots,x_n,y_1,\ldots,y_m)$ where B has no quantifiers. For arbitrary γ, and $1 \le r \le m$, there are functions $f_r(P;c_1, \ldots,c_n,c_1',\ldots,c_{r-1}')$ defined for c_i, c_j' in the union of all S_δ, $\delta < \alpha$, with the following property: If $Q_r = \exists$, f_r is the least γ such that P forces

(1) $$Q_{r+1}y_{r+1}\ldots Q_my_m B(c_1,\ldots,c_n,c_1',\ldots,c_{r-1}',y_{r+1},\ldots,y_m)$$

for some $c_\gamma \in S$, $f_r = 0$ if no such γ exists. Let $g_r(c_1,\ldots,c_n, c_1',\ldots,c_{r-1}') = \sup\{f_r | \text{all } P\}$. Again, by the previous section g_r is defined in M. If $Q_r = \forall$, f_r and g_r are defined the same way with (1) replaced by its negation. Let γ_1 be the supremum of the range of g_r for all r, $1 \le r \le m$, with c_i and c_i' running through the union of all S_δ, $\delta < \gamma$. Write $\gamma_1 = h(\gamma)$. Clearly h is defined in M. Now define $\beta_1 = \beta$ and $\beta_{n+1} = h(\beta_n)$, and $\beta' = \sup \beta_n$. It is then clear that our lemma holds if c is taken as the set defined by the formula "c is the union of $M_\gamma(a)$ for all $\gamma < \beta'$." This c corresponds to the set we have called $X_{\beta'}$, and our construction easily allows the formation of this set.

We have seen that the range z of φ on $\bar{c}_0$ is contained in some $\bar{c}_1$ and hence can be characterized as $\{\bar{c} \in \bar{c}_1 | \exists x\ (x \in \bar{c}_0\ \& A_n(x,\bar{c},\bar{c}_2,\ldots,\bar{c}_n))\}$ where $\bar{c}_2,\ldots,\bar{c}_n$ are the constants which occur in the definition of φ. By the theorem, we can assume that all the bound variables in A are restricted to lie in fixed sets of N. By the construction of N, this defines z by a formula of the type which implies $z \in N$ and the Replacement Axiom is proved. Thus N is a model for ZF.

LEMMA. For each $\alpha < \alpha_0$, $\exists\, c_\alpha$ such that rank $\bar{c}_\alpha = \alpha$ independently of which $\{P_n\}$ is taken as the complete sequence.

Proof. The lemma is obvious for $\alpha \leq \omega$. Now, rank $M_0(a) = \omega + 1$ and it follows by induction that rank $M_\alpha(a) = \omega + 1 + \alpha$.

From the set c_α with rank $\bar{c}_\alpha = \alpha$, one easily finds a c_β such that $\bar{c}_\beta = \alpha$ independently of the forcing sequence. This is a consequence of the following general principle: Let $x = R(y_1, \ldots, y_n$ be an absolute relation defining x as a function of the y_i. Suppose for each α, we have a β such that if $c_i \in \cup\{S_\gamma | \gamma \leq \alpha\}$, there is an element x in X_β such that $x = R(\bar{c}_1, \ldots, \bar{c}_n)$ holds when relativized to X_β where β is independent of $\{P_n\}$. Then for some $c \in S_\beta$, $\bar{c} = x$ independently of $\{P_n\}$. In the above case it is only a tedious repetition of the same ideas as in the proof of the absoluteness of the rank function to give β as a function of α. We shall use this principle again without explicit mention if it is obvious that we are dealing with absolute relations of the above type.

THEOREM. In N, $\bar{a}$ is not constructible.

Proof. Let c_α be an element of S such that $\bar{c}_\alpha = \alpha$ independently of $\{P_n\}$. Assume x is the element of L constructed at the α-th stage under any natural well-ordering of the construction. Since the construction is absolute α constructs x in N also. For any P since P is finite, $\exists$ Q and n such that $Q \supseteq P$ and either $(n \in a) \in Q$ and $\sim n \in a$ or $(\sim n \in a) \in Q$ and $n \in x$. If P forced "c_α constructs a" then α would have to construct $\bar{a}$ in the model N' defined by a complete sequence $\{Q_n\}$ where we can take $Q_0 = Q$. This is a contradiction. Thus since the ordinals in N are the same as those in M and $\bar{a}$ is never forced by any P to be constructible by any c_α, $\bar{a}$ is not constructible by any c_α, $\bar{a}$ is not constructible in N. Here we have used for the first time the fact that each P is a finite set.

THEOREM. GCH and AC hold in N.

Proof. The relation $x \in M_\alpha(\bar{a})$ can be expressed in ZF as an absolute relation $F(x, \alpha, \bar{a})$. This is seen by the same argument that showed that $x \in M_\alpha$ was expressible in ZF as an absolute relation. Since $M_0(\bar{a})$ is well-ordered, again the same argument as for L shows that one has an induced well-ordering on all of N expressible by a formula which has the single constant $\bar{a}$.

The proof of GCH goes over precisely as before again with the one modification that arguments about the relation "$x \in M_\alpha$" are replaced by "$x \in M_\alpha(\bar{a})$". We sketch the proof for the key lemma.

LEMMA. If $x \in M_\alpha(\bar{a})$, α infinite, and $y \subseteq x$ for some y in N, then $\exists \beta$ such that $\bar{\bar{\beta}} = \bar{\bar{\alpha}}$ and $y \in M_\beta(\bar{a})$.

<u>Proof</u>. A word of caution. The relation $\bar{\bar{\beta}} \leq \bar{\bar{\alpha}}$ can be interpreted in either M or N. Since $M \subseteq N$, it is possible that $\bar{\bar{\beta}} < \bar{\bar{\alpha}}$ holds in M but $\bar{\bar{\beta}} = \bar{\bar{\alpha}}$ in N. Actually, when we discuss CH we shall give an argument which when applied to this case shows that this cannot happen. In this case, the relation is to be interpreted in N.

Assume $y \in M_\gamma(\bar{a})$. Let A denote the set $\cup\{M_\beta(\bar{a}) \mid \beta \leq \alpha\} \cup \{y, \gamma\}$. Since AC holds in N we can apply the Löwenheim-Skolem theorem of Chapter II. Thus, there is a set B, $B \supseteq A$, $\bar{\bar{B}} = \bar{\bar{\alpha}}$ and such that $x \in M_\alpha(\bar{a})$ holds when relativized to B, and the Extensionality Axiom holds in B. Let φ map B isomorphically one-one onto a transitive set B'. Then φ is the identity on $\cup\{M_\beta(\bar{a}) \mid \beta \leq \alpha\}$ since this set is already transitive. Also $\varphi(y) = y$ and $\varphi(\gamma) = \beta$, where β is some ordinal and $\bar{\bar{\beta}} \leq \bar{\bar{\alpha}}$. By the usual absoluteness argument we get that $y \in M_\beta(\bar{a})$.

It is also possible to prove GCH in N by an argument which counts the conditions P which force various statements and which is in total analogy with the proof of the Power Set Axiom. We leave it as an exercise for the reader to give the details of this proof. We now summarize our results about the model N.

THEOREM. From ZF + SM it follows that there is a standard model for ZF in which AC and GCH hold and which contains an $\bar{a} \subseteq \omega$ such that $\bar{a}$ is not constructible. Thus GCH and AC do not impy $V = L$.

We close with a theorem that restates the defining property of a complete sequence without using the notion of forcing. Even if we had defined complete sequences in this way, however, to show that N is a model we would still have to go through the same steps.

<u>Definition</u>. Let A be the set of all pairs $\langle P, Q\rangle$ where P and Q are disjoint finite subsets of ω. We write $\langle P_1, Q_1\rangle < \langle P_2, Q_2\rangle$ if $P_1 \subseteq P_2$,

$Q_1 \subseteq Q_2$. A subset B of A is called <u>dense</u> if i) for all $x \in A$, $\exists y \in B$ and $x < y$, ii) $x \in B$, $x < y$ implies $y \in B$.

If $\{P_n\}$ is an increasing sequence of forcing conditions, write $\bar{a} = \text{Lim } P_n$ if $\bar{a} = \{n | \exists k\ (n \in a) \in P_k\}$.

THEOREM. If $\bar{a} \subseteq \omega$, $\exists$ a complete sequence $\{P_n\}$ with $\bar{a} = \text{Lim } P_n$ if and only if for every dense subset B of A lying in M, $\exists n$ such that if $P = \bar{a} \cap \{0,\ldots,n\}$ and $Q = \{0,\ldots,n\} - P$, then $\langle P,Q \rangle \in B$.

<u>Proof</u>. Since whether $\{P_n\}$ is a complete sequence depends only on $\text{Lim } P_n$, the theorem gives a characterization of complete sequences. It is clear from the definition of a dense set that if B is given then for any P, $\exists Q \supseteq P$ such that Q forces $\bar{a}$ to have the property of the theorem. This implies that if $\{P_n\}$ is complete $\bar{a}$ has that property. Conversely, if $\bar{a}$ satisfies our property let P_n be any sequence with $\bar{a} = \text{Lim } P_n$. Then, for any given statement let B be the set (which is in M) of forcing conditions which force either the statement or its negation. This is a dense set and hence by our condition some P_n forces either the statement or its negation.

7. THE GENERAL FORCING CONCEPT

The method of forcing is applicable to many problems in set theory. Rather than immediately proceeding to specific problems such as the independence of CH, we shall give a very general result which will cover the construction of all the models. We will always start with the model M and adjoin certain generic sets and then take N as the class of all sets generated from these using the ordinals of M. There is, in general, no obvious way to say how we are to generate sets from a given collection of sets since the "collecting" process can be performed at various stages. More precisely, consider the following problem. Let A be a given transitive set. We wish to find a transitive model N such that $A \subseteq N$ and if N' is another such model, $N \subseteq N'$. If we demand $A \in N$, the problem is trivial provided we drop AC since we just take $M_0 = A$, and $M_\alpha = (\bigcup_{\beta < \alpha} M_\beta)'$. We conjecture that for suitable A, no such N exists. The possible ways of generating N from the generic sets are quite varied.

Also one may sometimes even want to introduce a class (relative to M) of generic sets, rather than merely a set. The presentation we give here will be sufficient for the examples we have in mind, but it should be clear to the reader how it can be extended.

We assume given a "label" space $S = \cup\{S_\alpha | \alpha < \alpha_0\}$. A certain subset G of S are called generic sets. If $c \in S_\alpha - G$ then c corresponds to a formula $A(x)$ which contains as constants certain elements of $X_\alpha = \cup\{S_\beta | \beta < \alpha\}$ and all bound variables are restricted to X_α. For a fixed α_1 we assume that for $\alpha > \alpha_1$ this is a one-one map between these formulas and S_α. (The fact that for some α there are some formulas which do not correspond to any c will not affect our arguments.) All our maps and sets are assumed of course to lie in M. The general idea is that finally the sets we construct, with $c \in S_\alpha$, shall be composed of sets of the form $\bar{c}'$ with $c' \in S_\beta$ for some $\beta < \alpha$. We shall not be that careful in practice in that if the definition of some c is absolutely clear we shall allow it and all its members to occur in the same S_α. The definition of limited statements and rank is precisely as before.

Definition. A set of forcing conditions is a set U and a relation $<$ on U, both in M, such that if P, Q, R are in U then $P < P$ and $P < Q$ & $Q < R \to P < R$. Also, there is a map ψ in M such that if $P \in U$, $\psi(P)$ is a set of statements of the form $\{c_1 \in c_2\}$ where $c_2 \in G$, and if $c_2 \in S_\alpha$ then $c_1 \in S_\beta$ for some $\beta < \alpha$. If $P < Q$, then $\psi(P) \subseteq \psi(Q)$.

Definition. We define P forces a limited statement by induction in the rank as follows.

1-8. Exactly as in § 3.

9. P forces $c_1 \in c_2$ where $c_1 \in S_\alpha$, $c_2 \in S_\beta$, $\alpha < \beta$ if

   i) $\sim c_2 \in G$ and P forces $A(c_1)$ where $A(x)$ is the formula assigned to c_2

   ii) $c_2 \in G$ and for some $c_3 \in S_\gamma$, $\gamma < \beta$, $\{c_3 \in c_2\} \in \psi(P)$ and P forces $c_1 = c_3$.

10. P forces $c_1 \in c_2$ where $c_1 \in S_\alpha$ and $c_2 \in S_\beta$, $\alpha \geq \beta$ if for some $c_3 \in S_\gamma$, $\gamma < \beta$, P forces $\forall_\alpha x\,(x \in c_1 \leftrightarrow x \in c_3)$ & $(c_3 \in c_2)$.

The definition of forcing for unlimited statements is precisely as before. We have here insisted that only $\emptyset$ belong to S_0 and that for all α, an element of S_α has as its members only elements of $\cup\{S_\beta|\beta < \alpha\}$. No negative conditions $\sim c_1 \in c_2$ were included as direct consequences of any P, because our definition of forcing automatically implies $\sim c_1 \in c_2$ will be forced if $c_1 \in c_2$ is not forced. In our construction we have restricted ourselves in that we form sets corresponding to quantification over $X_\alpha = \cup\{S_\beta|\beta < \alpha\}$. This is a rather liberal definition in that there is no necessity to "collect" as often as the X_α do. If we examine our proof that N is a model, we use only two facts. The first is that for each α there is some c in S such that we always have $\bar{c} \supseteq S_\alpha$. This is used in the proof of the Power Set Axiom and the Replacement Axiom. The second property is that for any ordinal-valued function f in M, if α_0 is given, there is a sequence α_n such that $\alpha_{n+1} \geq f(\alpha_n)$ and there is a $c \in S$ such that $\bar{c} = \cup\{S_{\alpha_n}|n\}$. This is used in the Löwenheim-Skolem argument in the proof of the Replacement Axiom.

In our present situation, we can now prove exactly as before that all the main lemmas hold. The definition of a complete sequence $\{P_n\}$ is the same. To define the model N we proceed by induction on α. For $c \in S_0$, $\bar{c} = \emptyset$. If $\bar{c}$ has been defined for all $c \in S_\beta$, $\beta < \alpha$, then if $c \in S_\alpha - G$ we define $\bar{c}$ by the condition $A(x)$ which corresponds to c, exactly as before. If $c \in S_\alpha \cap G$, then $\bar{c} = \{\bar{c}_1 | \exists n \;\&\; (c_1 \in c_2) \in \psi(P_n)\}$. Again, the statements that are true in N are exactly those which are forced by some P_n. The proof that N is a model now carries over exactly as before. This is seen by examining the proof given in our previous case and seeing that no special properties were used other than the general features of forcing which hold in the new context also.

In most of the applications we make, it will always be that $P < Q \longleftrightarrow \psi(P) \subseteq \psi(Q)$. However, more subtle situations are possible. Thus, we might have $\psi(P) \not\subseteq \psi(Q)$ and $P = Q$. This means that although the immediate implications of P and Q are the same, they may have different effects because for some R, $P < R$ yet $\sim Q < R$.

8. THE CONTINUUM HYPOTHESIS

Let $\aleph_\tau$, $\tau \geq 2$, be a fixed cardinal in M. Let S be defined as follows: For all $\alpha < \aleph_\tau$, S_α consists of one element c_α and it will eventually turn out that $\bar{c}_\alpha = \alpha$. All these $c_\alpha \in G$. For $\alpha = \aleph_\tau$, S_α consists of $\aleph_\tau$ elements, all in G, which we denote by a_δ, $\delta < \aleph_\tau$. For these we will have $\bar{a}_\delta \subseteq \omega$ and their presence will guarantee that the continuum is at least $\aleph_\tau$. Still writing $\alpha = \aleph_\tau$, $S_{\alpha+1}$ consists only of elements of G as follows: There are $\aleph_\tau$ elements which will eventually be $\{\beta\}$ for all $\beta < \aleph_\tau$ and we denote them by the symbol $\{\beta\}$. There will also be $\aleph_\tau$ elements which eventually will become $\{\delta, a_\delta\}$ for $\delta < \aleph_\tau$ and we denote them by $\{\delta, a_\delta\}$. $S_{\alpha+2}$ consists of $\aleph_\tau$ elements in G which eventually become $\langle \delta, a_\delta \rangle$ for $\delta < \aleph_\tau$ and we denote them by $\langle \delta, a_\delta \rangle$. Finally $S_{\alpha+3}$ consists of one element in G, which we denote W and $\bar{W}$ will eventually be $\{\langle \delta, a_\delta \rangle | \delta < \aleph_\tau\}$. S_β for all $\beta > \alpha + 3$ contains no elements of G and its elements are in one-one correspondence with formulas ranging over $\cup\{S_\gamma | \gamma < \beta\}$. In this construction we are primarily interested in the a_δ and the set W, but since $\langle x,y \rangle = \{\{x\}, \{x,y\}\}$ we have to define W in a sequence of steps. The forcing conditions consist of all finite sets P of statements of the form $n \in a_\delta$ or $\sim n \in a_\delta$ for $n < \omega$, $\delta < \aleph_\tau$ which do not contain both $n \in a_\delta$ and $\sim n \in a_\delta$ for any n, δ. We write $P < Q$ if $P \subseteq Q$ and $\psi(P)$ consists of all the statements $c_n \in a_\delta$ where $(n \in a_\delta)$ is in P together with the following fixed statements:

i) $c_\alpha \in c_\beta$ if $\alpha < \beta < \aleph_\tau$.

ii) $c_\alpha \in \{\alpha\}$, $\delta \in \{\delta, a_\delta\}$, $a_\delta \in \{\delta, a_\delta\}$ where $\alpha, \delta < \aleph_\tau$.

iii) $\{\delta\} \in \langle \delta, a_\delta \rangle$ and $\{\delta, a_\delta\} \in \langle \delta, a_\delta \rangle$ for $\delta < \aleph_\tau$.

iv) $\langle \delta, a_\delta \rangle \in W$ for $\delta < \aleph_\tau$.

Let $\{P_n\}$ be a complete sequence and let N be the resulting model for ZF. Then $\bar{a}_\delta \subseteq \omega$ and clearly the a_δ are distinct since no P can force $a_\delta = a_{\delta'}$ if $\delta \neq \delta'$, as the P are merely finite sets and for any P, $\exists\, Q \supseteq P$ such that for some n $(n \in a_\delta)$ and $(\sim n \in a_\delta)$ are in Q.

LEMMA 1. In N, every set is constructible from $\bar{W}$. That is, if M_0 is the transitive closure of $\bar{W}$, and $M_\alpha = (\cup\{M_\beta | \beta < \alpha\})'$, then $N = \cup\{M_\alpha | \alpha \in M\}$.

<u>Proof</u>. Here transitive closure of X means $\bigcup_n X_n$ where $X_0 = X$ and X_{n+1} is the sum-set of X_n. The lemma is obvious since $\cup\{\bar{S}_\alpha | \alpha \leq \aleph_\tau + 3\}$ is precisely that closure.

THEOREM 1. AC holds in N.

Proof. Since the construction $X \to X'$ is absolute we can express in N the relation $x = M_\alpha$ of Lemma 1. Since M_0 is well-ordered, because of $\bar{W}$, this induces, by the usual arguments, well-orderings of all M_α and hence of N. The well ordering is expressible by a formula which has one one constant, $\bar{W}$.

Definition. If there is no R such that $P < R$ and $Q < R$, we say P and Q are incompatible.

LEMMA 2. Let B be a set in M of mutually incompatible P, then B is countable in M.

Proof. The entire argument takes place in M. Assume B is uncountable. Let B_n be the subset of B consisting of those P which contain less than n statements. Some B_n must be uncountable (since AC holds in M so that a countable union of countable sets is countable. Actually the set of all P is well-ordered so we do not need AC.). Thus we may assume all P in B have less than n statements. Let k be the largest integer such that for some P_0 with k statements there are uncountably many P in B such that $P_0 < P$. We may have $k = 0$ and $P_0 = \emptyset$. In any event, cut down B so that it consists only of such P. Let P_1 be an arbitrary element of B and let $A_1, \ldots, A_m$ be the statements in $P_1 - P_0$. Since P_1 is incompatible with all other P in B, and all P in B contain P_0, it must follow that for some A_i, say A_1, uncountably many P in B contain $\sim A_1$. Now, $\sim A_1$ is not in P_0 since otherwise $\sim A_1$ would be in P_1, and there are uncountably many P in B which contain $P_0 \cup \{\sim A_1\}$ which contradicts the definition of k.

LEMMA 3. Let f be a single-valued function definable in N. There is a function g, definable in M, which assigns to each c in S a countable subset, g(c), of S and such that for all c, $f(\bar{c}) = \bar{c}'$ for some $c' \in g(c)$.

Proof. The point of this lemma is that although we cannot describe f in M, we can be sure that the range of f is sufficiently limited so that we will be able to prove in the next theorem that cardinalities are not confused. We mean that f is defined by a sentence which may involve $\bar{c}$ for some c in S. We can thus form statements in our formal language referring to f. Until further notice, all our notions are in M. For each c, c' in S, let $A(c,c')$ be the set of all P such that P forces f to be single-valued and c' is the first element of S (under a natural well-ordering), such that for any $Q > P$, Q forces $f(c) = c'$. Clearly if $c' \neq c''$, the elements of $A(c,c')$ are incompatible with those of $A(c,c'')$. Therefore, using AC in M and Lemma 2 there are only countably many c' such that $A(c,c') \neq \varphi$. Let $g(c)$ be the set of these c'. We return now to N, let c' be the first element of S such that $f(\bar{c}) = \bar{c}'$. Since the construction of S_α can be expressed in N, and the well-ordering of S expressed in N, we can consider as a legitimate unlimited statement, $T \equiv$ "c' is the first element of S such that $f(c) = c'$ ". Some P_n must then force T and the fact that f is single-valued since these are true in N. Hence $A(c,c')$ is not empty and $c' \in g(c)$ and the lemma is proved.

THEOREM 2. Let α, β be ordinals $\bar{\bar{\alpha}} < \bar{\bar{\beta}}$ in M. Then $\bar{\bar{\alpha}} < \bar{\bar{\beta}}$ in N.

<u>Proof</u>. We can clearly assume that α and β are cardinals. Let f be a single-valued function in N which maps α onto β such that $f(x) = 0$ if $\sim x \in \alpha$. Put $X_\gamma = \bigcup\{S_\delta | \delta < \gamma\}$. Clearly $\gamma \in X_{\gamma+3}$ for all γ and $x \in X_\gamma \to \text{rank } x \leq \gamma$. By the proof of Lemma 3, it follows that the range of f on $X_{\alpha+3}$ is contained in $\bar{T}$ where T is a subset of X_β and the cardinality of T in M is $\leq \aleph_0 \cdot \bar{\bar{\alpha}} = \bar{\bar{\alpha}}$. $\bar{T}$ contains the range of f. If $c \in T$, then $c \in S_\gamma$ for some $\gamma < \beta$. Since the cardinality of T is $\leq \alpha$ in M, it follows that since $\bar{\bar{\alpha}} < \bar{\bar{\beta}}$ in M, there is a $\gamma < \beta$ such that $T \subseteq \bigcup\{S_\delta | \delta < \gamma\}$. But then rank $\bar{T} \leq \gamma$ and so $\bar{T}$ cannot contain β and thus the range of f cannot include all of β.

We now know that the cardinals in M are precisely the cardinals in N. Since the a_δ are all distinct we see

THEOREM 3. In N, $C \geq \aleph_\tau$. Thus CH is false in N.

Having shown that $C \geq \aleph_\tau$, the problem remains to determine the cardinality of C in N precisely. We give two classical results.

LEMMA 4 (König). Let A_α and B_α be collections of sets such that $\bar{\bar{A}}_\alpha < \bar{\bar{B}}_\alpha$ for all α. Then $\overline{\overline{\cup A_\alpha}} < \overline{\overline{\Pi B_\alpha}}$. Here Π denotes the direct product.

Proof. Assume that f is a function from $\cup A_\alpha$ onto ΠB_α. Let f_β be the projection of f on B_β. Then for each α, f_α cannot map A_α onto B_α. Choose $x_\alpha \in B_\alpha$ not in the range of f_α on A_α. Then $\prod_\alpha x_\alpha$ is not in the range of f.

LEMMA 5. The continuum C is not the sum of countably many smaller cardinals.

Proof. This is the only known restriction on C. If $\bar{\bar{A}}_n < C$, then $\Sigma \bar{\bar{A}}_n < \prod_{n=1}^{\infty} C = (2^{\aleph_0})^{\aleph_0} = 2^{\aleph_0} = C$. We shall show that in some sense this is the only restriction possible on C.

LEMMA 6. In M (i.e., assuming GCH), the number of countable subsets of $\aleph_\tau$ is $\aleph_\tau$ if $\aleph_\tau$ is not a countable sum of smaller cardinals, $\aleph_{\tau+1}$ if it is.

Proof. Observe first that the number of countable subsets of $\aleph_\delta$ for any δ is $\leq \aleph_\delta^{\aleph_0} \leq 2^{\aleph_\delta \cdot \aleph_0} = \aleph_{\delta+1}$ by GCH. Now assume that $\aleph_\tau$ is not a countable sum of smaller cardinals. This means that there is no sequence τ_n, $\tau_n < \tau$ and yet $\sup \tau_n = \tau$. We say then that τ is not cofinal with ω. Now, any countable subset of $\aleph_\tau$ is thus contained in $\aleph_{\tau'}$, for some $\tau' < \tau$. The number of such sets is $< \aleph_{\tau'+1} \leq \aleph_\tau$. Thus the number of countable subsets of $\aleph_\tau$ is $\leq \bar{\bar{\tau}} \cdot \aleph_\tau = \aleph_\tau$. On the other hand, assume now that $\aleph_\tau = \Sigma A_n$ where A_n are cardinals and $A_n < \aleph_\tau$. Assume $A_n < A_{n+1}$ and put $B_n = A_n - A_{n-1}$. The number of countable subsets is at least $\Pi\bar{\bar{B}}_n = \Pi A_n$. But $\aleph_\tau = \Sigma A_n < \Pi A_{n+1}$ and so the number is at least $\aleph_{\tau+1}$, but is also at most $\aleph_{\tau+1}$, hence is equal to it.

Definition. For any statement A, we say that P is minimal for A, if P forces $\sim\sim A$ (i.e., no $Q > P$ forces $\sim A$), and no $P' < P$ has this property.

To say that P forces $\sim\sim A$ is precisely the condition that if $P < P_n$ where P_n is one of the terms of a complete sequence defining N, then A holds in N.

LEMMA 7. For any statement A, the set of minimal P for A is countable in M.

Proof. This proof is almost identical with that of Lemma 2. Let B be an uncountable set of minimal P. As in Lemma 2, we can assume that all the P in B contain n statements. Let k be the largest integer such that there is a P' with k conditions and $P' < P$ for uncountably many P in B. We can then assume $P' < P$, for all P in B. Now since the P in B are minimal, P' cannot force $\sim\sim A$. Hence, for some $P'' > P'$, P'' forces $\sim A$ and hence P'' is incompatible with all P in B. As in Lemma 2, this means that for some $c \notin P'$, $P' \cup \{c\}$ is contained in uncountably many P in B which contradicts the definition of k.

We can now determine the cardinality of C.

THEOREM 4. In N, $C = \aleph_\tau$ if τ is not co-final with ω in M, $C = \aleph_{\tau+1}$ if τ is co-final with ω.

Proof. For each $c \in S$ and $n \in \omega$, let $V(n,c)$ be the set of minimal P for $n \in c$. If $V(n,c) = V(n,c')$ for all n and $\bar{c} \subseteq \omega$, $\bar{c}' \subseteq \omega$, then $\bar{c} = \bar{c}'$. This is true because if P_k is in the complete sequence and P_k forces $n \in c$, there must be a minimal P', $P' < P_k$. Then P' and hence P_k force $\sim\sim n \in c'$ so $n \in \bar{c}'$. Thus $\bar{c} = \bar{c}'$. For fixed n, the number of possible sets $V(n,c)$ is at most $\bar{\bar{D}}$ = the number of countable subsets of $\aleph_\tau$, as computed in M. Now we claim that if E is the set of countable sequences in D, $\bar{\bar{E}} = \bar{\bar{D}}$. (Again, all notions here are in M.) For, each element of E gives rise to a countable subset of D and each subset comes from at most $2^{\aleph_0}$ sequences, so that by Lemma 6, we get $\bar{\bar{E}} \leq \bar{\bar{D}} \cdot \aleph_1 = \bar{\bar{D}}$, and since $\bar{\bar{D}} \leq \bar{\bar{E}}$, $\bar{\bar{D}} = \bar{\bar{E}}$. We have thus shown that in N, $C \leq \bar{\bar{D}}$. If τ is co-final with ω, we see that in N, $\aleph_\tau \leq C \leq \aleph_{\tau+1}$. Lemma 5 says that C cannot equal $\aleph_\tau$ so $C = \aleph_{\tau+1}$. If τ is not co-final with ω, we have $C \leq \aleph_\tau$ and since $C \geq \aleph_\tau$, $C = \aleph_\tau$.

It follows now that we can construct models in which $C = \aleph_2$, $\aleph_{\omega+1}$, $\aleph_{\omega^2+1}$, etc. We can even have $C = \aleph_{\omega_1}$ where $\omega_1 = \aleph_1$. To see this, we note that ω_1 has the same meaning in M as in N since cardinals do not change, and for the same reason $\aleph_{\omega_1}$ also retains its meaning. We can give a precise statement as follows: Consider a formula $A(x)$ which has the property that one can prove that in any transitive set or class which satisfies ZF, $A(x)$ defines a unique ordinal. Assume further that $A(x)$ is absolute in the sense that if S_1 and S_2 are two such classes, $S_1 \subseteq S_2$, for which the notion of cardinal is the same, then $A(x)$ defines the same element in S_1 and S_2. It is then possible to give a model for ZF in which, if $A(\alpha)$ holds for some α, then if α is not co-final with ω, then $C = \alpha$. The details are left to the reader.

In [4], the determination of the cardinality of C in N used a more straightforward but less elegant construction. We give a rough sketch. Consider the proof that in L, every real number is constructible by a countable ordinal. The essential idea there was that if $x \subseteq \omega$ and $x \in M_\alpha$, then by taking the Skolem "hull" of α, ω, and x, we can get an isomorphism of $\alpha \to \alpha'$ where α' is countable and $x \in M_{\alpha'}$. In N, every set is constructible from $W = \{\langle\delta, a_\delta\rangle \mid \delta < \aleph_\tau\}$. We would like to do the computation of the cardinality of this hull in M, since in M we know the arithmetic of cardinals very well. In the course of constructing the hull we have to perform countably many choices in N. However, by considering all possible P which force such choices, one can show that a hull can be constructed by performing countably many choices, and the set of $c \in S$ which are picked out form a set in M and this set is countable. Since we start with W in the hull, we may have to choose certain elements of W. It turns out that the hull we construct completely determines x. Since all the hulls have only countably many elements, the crucial point is which countable subsets of W are contained in it. In this way one can see that the cardinality of C is not greater than that of the set of all countable subsets of $\aleph_\tau$.

Results have been obtained [6], [25], yielding models in which various simultaneous statements hold in cardinal arithmetic.

In [6], it is proved that if $G(\alpha)$ is any function in M such that (i) $\alpha \leq \beta$ implies $G(\alpha) \leq G(\beta)$ and (ii) $\aleph_{G(\alpha)}$ is not co-final with any cardinal $\leq \aleph_\alpha$, then in N, $2^{\aleph_\alpha} = \aleph_{g(\alpha)}$ for all regular cardinals $\aleph_\alpha$. (A regular cardinal is one which is not co-final with a smaller cardinal.) A further restriction on cardinal equalities is clearly that if $2^{\aleph_\beta} = \aleph_\gamma$ for all β such that $\delta \leq \beta < \alpha$ where δ, α are fixed and α is singular (not regular), then $2^{\aleph_\alpha} = \aleph_\gamma$. The techniques involved in [6] consist of introducing generic sets at different levels, although care must be taken that generic sets introduced on one level do not affect the cardinalities of a much lower level but only those cardinals that they were intended to affect.

9. THE AXIOM OF CHOICE

We next discuss models N in which AC fails. Classical results in this direction were obtained by Frankel and Mostowski who dropped the Axiom of Extensionality and introduced "atoms", i.e., fictitious objects x_i such that $\forall y\, (\sim y \in x_i)$ yet $x_i \neq x_j$ for $i \neq j$. Since such atoms x_i are certainly quite difficult to distinguish, it is possible to introduce certain symmetries in models of set theory. For example, if we introduce atoms x_n, y_n we can arrange things so that every set is invariant under all the transpositions $(x_n y_n)$, except for finitely many n. Clearly one must allow for such exceptions since x_1, say, is not invariant under $(x_1 y_1)$. In our models generic sets will play a role similar to these atoms. However, for a standard model N, there can never be a true automorphism of N since if σ were such an automorphism, one proves by induction on rank x, that $\sigma x = x$. This complicates matters, but the basic idea of having some kind of symmetry remains.

Results have also been obtained by dropping the Regularity Axiom and keeping Extensionality. In this case the role of an atom is played typically by a set x_0 such that $x_0 = \{x_1\}, \ldots , x_n = \{x_{n+1}\}, \ldots$. These results, of course, imply nothing about well-founded sets, e.g., whether the real numbers can be well-ordered.

In our first model, we shall introduce a set V which consists of infinitely many generic subsets of ω. The construction will be so arranged that one can only distinguish between finitely many of them with any one set. Thus, S_0 consists of all integers. S_1 consists of countably many symbols a_n, each of which is generic and will eventually be a subset of ω. S_2 consists of one generic set V which will eventually have as its members precisely the sets a_n. S_α for $\alpha > 2$ consists of labels for <u>all</u> formulas in ZF relativized to $\cup\{S_\beta | \beta < \alpha\}$ possibly involving constants from that set. The forcing conditions P consist of all finite collections of statements of the form $n \in a_m$ or $\sim n \in a_m$

which do not contain both a statement and its negation. The partial ordering on the sets P is just inclusion. For each P, $\psi(P)$, the set of elementary statements forced by P, consists of the following

i) if $m, n \in S_0$ and $m < n$ then $(m \in n) \in \psi(P)$

ii) if $(m \in a_n) \in P$ then $(m \in a_n) \in \psi(P)$

iii) $(a_n \in V) \in \psi(P)$ for all n.

Let G be the group of all permutations π of ω such that $\pi(n) \neq n$ for only finitely many n, and let G_n be the subgroup of G equal to $\{\pi | \pi(m) = m \text{ for } m \leq n\}$. We shall now define how G acts on $S = \bigcup_\alpha S_\alpha$.

<u>Definition</u>. For $c \in S_0$, $\pi \in G$, let $\pi c = c$. For $a_n \in S$, let $\pi(a_n) = a_{\pi(n)}$. Let $\pi(V) = V$. For $\alpha > 2$, $c \in S_\alpha$, assume c corresponds to a formula $A(x, c_1, \ldots, c_m)$ where $c_i \in X_\alpha = \bigcup\{S_\beta | \beta < \alpha\}$ and A is understood to have all variables restricted to X_α. Then $\pi(c)$ is the element of S_α corresponding to $A(x, \pi(c_1), \ldots, \pi(c_m))$.

<u>Definition</u>. Let A be a limited statement, $A \equiv B(c_1, \ldots, c_n)$ where $c_i \in S$ and B is a formula of ZF such that each quantifier has an ordinal subscript, i.e., $\forall_\alpha$, $\exists_\alpha$. Then $\pi(A) \equiv B(\pi(c_1), \ldots, \pi(c_n))$, where the ordinal subscripts are left unchanged. Similarly for unlimited statements.

<u>Definition</u>. If P is a forcing condition, $\pi(P)$ is the forcing condition defined by $(n \in a_m) \in P \longleftrightarrow (n \in a_{\pi(m)}) \in \pi(P)$ and $\sim(n \in a_m) \in P \longleftrightarrow \sim(n \in a_{\pi(m)}) \in \pi(P)$.

LEMMA. P forces $A \longleftrightarrow \pi(P)$ forces $\pi(A)$.

<u>Proof</u>. This lemma expresses the sense in which the elements of G are automorphisms of our system. Let A be a limited statement. The proof proceeds by induction on rank A. It is obvious that the lemma need only be checked for limited statements A such that rank $A = (\alpha, i, r)$ and $\alpha \leq 2$ since the induction is trivial. Here we quickly reduce to the case A is of the form $n \in a_m$. But then it is obvious from our definition of the action of G on P and A. The result for unlimited statements follows by an induction on the number of symbols.

LEMMA. For each $c \in S$, statement A, and condition P, $\exists m$ such that $\pi \in G_m \to \pi(c) = c$, $\pi(A) = A$, $\pi(P) = P$.

Proof. If $c \in S_\alpha$ and corresponds to a formula $A(x, c_1, \ldots, c_n)$ such that for some m_i, $\pi \in G_{m_i} \to \pi(c_i) = c_i$, then if $m = \max(m_i)$ clearly $\pi \in G_m \to \pi(c) = c$. If $c \in S_0$, or $c \in S_2$, $\pi(c) = c$ for all π. If $a_m \in S_1$, then $\pi(a_m) = a_m$ if $\pi \in G_m$. The case of statements is handled similarly. The case of conditions P is trivial. Thus, although there are infinitely many a_n, each c is symmetric with respect to all of them with possibly finitely many exceptions.

Let N now denote the model obtained by choosing a complete sequence $\{P_n\}$.

THEOREM 1. N is a model for ZF, in which the set T is a subset of $P(\omega) = C$, such that T is infinite and yet contains no countable subset.

Proof. If $m \neq n$, then no P can force $a_m = a_n$ since we can always find some $Q > P$ which forces, for some k, $k \in a_m$ and $\sim k \in a_n$. Thus $\bar{a}_m \neq \bar{a}_n$ and clearly T is infinite. Let $c \in S$ and assume that some P of the complete sequence $\{P_n\}$ forces "c is a function f from ω into V, such that $f(m) \neq f(n)$ if $m \neq n$". Let r be such that $\pi \in G_r \to \pi(c) = c$ and such that r is greater than any m such that a_m occurs in a statement of P. In the complete sequence $\{P_n\}$ there must be some $P' \supseteq P$ such that P' forces $f(k) = a_s$ for some k and s where s is greater than r, since $\bar{f}$ takes infinitely many distinct values. Let $t > r$ be an integer such that a_t does not appear in P', and let π be the permutation which interchanges s and t and is the identity on all other integers. If $P'' = \pi(P')$, then P'' forces $f(k) = a_t$. Also, P' and P'' are compatible, i.e., $Q = P' \cup P''$ is a forcing condition, since P' and P'' are identical except for conditions involving a_s and a_t and P' does not involve t and P'' does not involve s. Thus Q forces f to be single-valued, since $Q \supseteq P$, and yet forces $f(k) = a_s$ and $f(k) = a_t$ which is impossible.

COROLLARY. In N the continuum is not well-ordered.

COROLLARY. In N, there exists a countable sequence $\{B_n\}$, each B_n a set of real numbers such that there is no function g such that $g(n) \in B_n$. Thus, the countable AC fails.

Proof. The real numbers are of course in one-one correspondence with subsets of ω so that C can be regarded either as the set of real numbers or $P(\omega)$. It is also well-known that for any n, there is a simple map of all n-tuples of reals one-one into the reals. (One "meshes" the numbers.) Let B_n be the set of reals corresponding to all n-tuples of elements of V which have no two members of the n-tuple equal. If g exists then one would obtain a countable subset of V.

Definition. We say P weakly forces A if P forces $\sim\sim A$.

Thus, if P weakly forces A, then if P is part of the complete sequence $\{P_n\}$, A must be true in N, since no $Q \supseteq P$ can force $\sim A$. Clearly if P forces A, P weakly forces A.

LEMMA. In N, if $x \in N$, $x \subseteq \omega$, there are finitely many $b_1 = a_{i_1}, \ldots, b_m = a_{i_m}$ such that x is constructible from $b_1, \ldots, b_m$.

Proof. We say x is constructible from $b_1, \ldots, b_m$ if when we define $M_0 = \omega \cup \{b_1, \ldots, b_m\}$, $M_\alpha = (\bigcup_{\beta<\alpha} M_\beta)'$, for some α, $x \in M_\alpha$. Let $x = \bar{c}$ and assume G_m keeps c fixed. For every P, let $\bar{P}$ denote the set of conditions in P which only involve a_i with $i \leq m$. We claim that if P forces $n \in c$, then $\bar{P}$ weakly forces $n \in c$.

If not, for some $P' \supseteq \bar{P}$, P' forces $\sim n \in c$. Clearly there is some $\pi \in G_m$ such that $\pi(P')$ and P are compatible. Then $\pi(P')$ forces $\sim n \in c$ and hence $Q = P \cup \pi(P')$ forces both $n \in c$ and $\sim n \in c$, impossible. The function h which sends any $\bar{P}$ into the set of all n such that $\bar{P}$ weakly forces $n \in c$ is expressible in M. Clearly x is now easily constructed from h and $a_1, \ldots, a_m$ since these a_i determine which $\bar{P}$ occur in the complete sequence.

The above lemma generalizes to any set x which is a subset of an ordinal. We have included it merely to show how the a_i behave independently. However, we need a slightly different form for an application.

To each element c of the label space S we shall assign a finite set of the a_n, the a_n upon which it "essentially" depends. The definition is obvious for $c \in S_0$, or $c \in S_1$. The element V is assigned $\emptyset$. Now if $c \in S_\alpha$ and corresponds to a formula which involves $c_1, \ldots, c_n$ where $c_i \in \bigcup\{S_\beta | \beta < \alpha\}$, we assign to c the set $\varphi(c)$ which is the union of the finite subsets $\varphi(c_i)$ of V which are assigned to the c_i. If B is a finite subset of V, then we consider all the elements c such that $\varphi(c) = B$ and call this class $D(B)$. By induction on α, we see that we can define on $S_\alpha \cap D(B)$ a natural well-ordering. Letting $\bar{D}(B)$ denote the corresponding sets in N, we see that $\bar{D}(B)$ is a class which is definable in terms of V with a well-ordering definable in terms of V. This is true because the construction we have used to define N can be described in N in terms of the one set V. Now, N is the union of all the $\bar{D}(B)$. If the sets B could be well-ordered we would have a well-ordering on N. This is of course not the case. However, the natural ordering of real numbers induces an ordering of the B, and we are tempted to conclude that we can define an ordering on all of N. To do this we need a lemma.

LEMMA. For each x in N, there is a B such that $x \in \bar{D}(B)$ and if $x \in \bar{D}(B')$ then $B \subseteq B'$.

We sketch the proof. Let $c_1 \in D(B_1)$, $c_2 \in D(B_2)$ and assume P_0 forces $c_1 = c_2$. Let $B = B_1 \cap B_2$. We shall show that for some $c_3 \in D(B)$, P_0 weakly forces $c_1 = c_3$. We only consider for the rest of this proof P such that $P \supseteq P_0$. Let $\bar{P}$ denote the part of P which involves $B_1 \cap B_2$. If $c' \in D(B')$ and P forces $c' \in c_1$, the same proof as in the previous lemma shows that if Q is the union of $\bar{P}$, P_0, and the part of P which involves B', that Q weakly forces $c' \in c_1$. We can give another definition, c_3, of the set $\bar{c}_1$. Namely, it consists of all $\bar{c}'$, where $c' \in \bigcup\{S_\beta | \beta < \alpha\}$ for suitable α, satisfying the following condition: If $c' \in D(B')$, let $F(c')$ be all forcing conditions Q which are the union of conditions involving only $B' \cup B$ and P_0 and such that Q weakly forces $c' \in c_1$. Then c_3, consists of the union taken over all B', of $c' \in D(B')$ such that some member of $F(c')$ is contained in an element of the complete sequence defining N. One can verify that this definition puts c_3 in $D(B)$. It is left as an exercise for the reader to verify this point.

Now, for each x in N, let B be the minimal finite subset of V such that $x \in D(B)$. All x corresponding to the same B have a definable well-ordering. Since the map from x to B is now single-valued, the ordering on the sets B induces a definable null-ordering in N. The result is the following theorem due to Lévy and Halpern [17].

THEOREM. In N there is a definable ordering of all sets which has only the set V occurring as a constant.

Thus, the fact that every set can be ordered does not imply that the continuum can be well-ordered.

Using similar ideas one can prove

THEOREM. If A is a set of reals in N which can be well-ordered, then for some n all the elements of A are constructible from $a_1, \ldots, a_n$.

A remark which follows from our proof is that every set can not only be ordered, but is in one-one correspondence with a subset of $\alpha \times C$ where α is some ordinal and C is the continuum. Another result concerning the relative strength of various consequences of AC is due to Halpern and Läuchli (unpublished as of now), and says that AC does not follow from the Prime Ideal Theorem. (The latter says that every Boolean algebra contains a non-trivial prime ideal.)

The following result is put here for lack of a better place.

THEOREM. Let N denote the model obtained in § 6 by adjoining one generic set $a \subseteq \omega$. Then, although AC holds no formula which does not use constants in N defines a well-ordering.

<u>Proof</u>. We sketch the proof [7]. Let a be the limit of a complete sequence. If for some a', $(a' - a) \cup (a - a')$ is only a finite set, then a' is again the limit of a complete sequence. For, given any statement A, the statement that some initial segment of a' forces either A or $\sim A$ is itself a statement which must be forced by some initial segment of a which clearly implies that a' is generic. Let R be the set of all subsets of ω which differ from a with respect to only finitely many integers. If A is a formula which well-orders N, let a' be the element of R that it picks out, and assume P forces a' to be picked out. Let a_0 be an element of R that agrees with a in all the integers

mentioned in P, yet $a_0 \neq a$, and let a_0' differ from a_0 in the same way a' differs from a. Then the above remark shows that by changing the complete sequence so that a is replaced by a_0, the model N, the set R and the formula A are unchanged. Yet now a_0' is picked out by A, a contradiction

We give an example to show that the countable AC may fail even for pairs. Let S_0 be the set of integers, S_1 be generic sets a_{ij}, b_{ij}, which will eventually be subsets of ω. S_2 consists of sets U_n, V_n, such that eventually $U_n = \{a_{n1}, a_{n2}, \ldots\}$, $V_n = \{b_{n1}, b_{n2}, \ldots\}$. S_3 consists of the symbols $\{U_n, V_n\}$, S_4 consists of the symbols $\{n, \{U_n, V_n\}\}$, and S_5 is the one set symbol V such that eventually V has its members exactly the sets of S_4. The forcing conditions consist of finitely many, non-contradictory, conditions of the form $n \in a_{ij}$, $n \in b_{ij}$, $\sim n \in a_{ij}$ or $\sim n \in b_{ij}$, as well as the obvious conditions to insure that the elements of $S_2, \ldots, S_5$ are as we have described them. Let G be the group of those permutations of a_{ij}, b_{ij} such that

i) $\pi(a_{ij}) = a_{ij}$, $\pi(b_{ij}) = b_{ij}$ for all $i > n$ for some n.

ii) for each i, either π maps the set U_i onto U_i and V_i onto V_i or else maps U_i onto V_i and V_i onto U_i.

Let G_m be the group of all π such that $\pi(a_{ij}) = a_{ij}$, $\pi(b_{ij}) = b_{ij}$ if $i \leq m$. One now reasons exactly as before to show that any c in S can only distinguish between finitely many U_i, V_i. If N denotes the model obtained from a complete sequence we have

THEOREM. In N, the countable AC fails for pairs of elements of $P(P(\omega))$.

Many other results concerning the relationship of various forms of AC can be attacked by these methods, and the reader is referred to the literature for the most recent results. One of the most interesting results is due to R. Solovay (as yet unpublished) which says that models N can be constructed in which the countable AC holds and yet every set of real numbers is Lebesgue measurable. His construction uses the idea of changing cardinalities which will be presented in the next section.

10. CHANGING CARDINALITIES

In this section we show how to produce a rather different phenomenon in the model N. Let α_0 be a fixed infinite ordinal in M. Let S_0 consists of all pairs $\langle n,\alpha\rangle$ where n is an integer and $\alpha < \alpha_0$, as well as the obvious sets required to make S_0 transitive. S_1 will consist of a single generic set T and S_α for $\alpha > 1$ are defined in the usual way by means of formulas quantifying over previous S_β, $\beta < \alpha$. A forcing condition P now consists of finitely many pairs of the form $\langle n_i,\alpha_i\rangle$ where $\alpha_i < \alpha_0$ and $n_i \neq n_j$, $\alpha_i \neq \alpha_j$ if $i = j$. Such a P forces precisely the sets $\langle n_i,\alpha_i\rangle$ to be in T. Let N be the resulting model. Clearly T will then give a one-one map of ω into α_0. Actually, the map is onto since for a given $\beta < \alpha_0$ clearly no P can force β not to be in the range of the map. Thus in N, α_0 becomes countable. If we do this process with $\alpha_0 = \aleph_1$, one can prove that $\aleph_1$ (in M) becomes countable in N, and in general $\aleph_{n+1}$ in M becomes $\aleph_n$ in N. (The latter is proved by imitating the proof of § 8 that cardinalities in that case were unchanged, but observing that now there can be $\aleph_1$ mutually incompatible forcing conditions.) As always in our construction no new ordinals are introduced in N and hence no new constructible sets. If we take $\alpha_0 = \aleph_1$, we get

THEOREM. There is a model N in which there are only countably many constructible real numbers.

Proof. Every constructible real is constructed by an ordinal α with $\alpha < \aleph_1$ in M.

This process of making cardinals countable was carried further by Lévy and Feferman [8], who showed

THEOREM. There is a model N such that in N, $\aleph_1$ is the supremum of countably many countable ordinals. Also the continuum is the union of a countable number of countable sets.

Proof. We give a sketch of the rather involved proof. The idea is that all the cardinals $\aleph_n$ will be made countable and $\aleph_\omega$ will become the new $\aleph_1$. Let S_0 be the set of all pairs $\langle n,\alpha\rangle$, $\alpha < \aleph_\omega$ together with the obvious sets needed to make S_0 transitive. We would like now to list sets which make $\aleph_n$ countable. If we merely did this and then started our usual process we would then have in N a set which gives a one-one correspondence of $\aleph_\omega$ with $\aleph_0$. To avoid this we proceed as follows. Let H be the group of all permutations of ω, which leave all but finitely many integers fixed, and let e be the identity of H. S_n for $1 \leq n < \omega$ consists of labels for generic sets which we denote πA_n where π ranges through H. Eventually each set πA_n will be a one-one map $f_{\pi,n}$ of ω onto $\aleph_n$ such that $f_{\pi,n}(k) = f_{e,n}(\pi k)$. Thus each S_n contains a whole collection of one-one maps of ω onto $\aleph_n$ and all of them will occur so symmetrically, that it will be impossible to pick out one for each n. We shall write A_n in place of eA_n. For the sake of the second part of the theorem we have to do one more step before we start our normal collecting process. Let S_ω consist of labels for the following sets:

i) for each n, $B_n = \{\pi A_n | \pi \in H\}$

ii) $T = \{\langle 1,B_1\rangle, \langle 2,B_2\rangle, \ldots\}$ and the other obvious sets necessary to make S_ω transitive.

For $\alpha > \omega$, we define S_α as all labels for formulas quantifying over previous S_β, $\beta < \alpha$. Each forcing condition P consists of a finite sequence $Q_1,\ldots,Q_n$ where each Q_i is a finite set of pairs $\langle m_j,\alpha_j\rangle$ where $\alpha_j < \aleph_i$ and $m_j \neq m_k$, $\alpha_j \neq \alpha_k$ if $j \neq k$. Such a P forces $\langle m_j,\alpha_j\rangle$ to belong to eA_i, and $\langle \pi m_j,\alpha_j\rangle$ to belong to πA_i, for all i and j. P also forces all the other obvious relations between the elements of S_α for $\alpha \leq \omega$. To illustrate the symmetry properties we define G as the weak direct product of ω copies of H, i.e., if $\pi \in G$, $\pi = \{\pi_1,\ldots\}$ where $\pi_i \in H$ and for some n, $\pi_i = e$ for $i > n$. If $\pi \in G$, $\pi = \{\pi_1,\ldots\}$ π acts on S as follows: $\pi(\pi' A_i) = \pi_i \pi' A_i$ and the action of G is extended in the obvious manner to all of S. G keeps each B_n fixed. Clearly G keeps each set S_α invariant. We let G_m denote the group of all $\pi = \{\pi_1,\ldots\}$ with $\pi_i = e$ for $i \leq m$. Again we have that for each c there is an m such that G_m keeps c fixed. G acts on the conditions P by sending each $\langle k,\alpha\rangle$ in Q_i into $\langle \pi_i k,\alpha\rangle$. G also operates on statements in the obvious way and again we have that P forces A if πP forces πA.

It is clear that if we define the rank of $X_\alpha = \cup\{S_\beta | \beta < \alpha\}$ by a suitable c, we can obtain for each α, a $c \in S$ such that $\bar{c} = \alpha$ independently of the complete sequence. Also c is invariant under G. Since each y in M is constructed by some α, we also see that we can obtain a c such that $\bar{c} = y$ independently of the complete sequence, and G leaves c invariant. This is a consequence of the general principle already mentioned that absolute constructions can be given by some c provided we have a bound for an α such that the construction can be carried out in some X_α.

LEMMA. If $x \in N$ and all the members of x are in M, then for some n, x is constructible from $A_1, \ldots, A_n$.

Proof. Let $x = \bar{c}_0$ and assume G_m keeps c_0 fixed. Then for each $y \in M$, there is a c invariant under G and $\bar{c} = y$. If any P forces $c \in c_0$ one can show that the part of P dealing only with A_i, $i \leq n$, already weakly forces $c \in c_0$. For, if not, then by applying G_m we could find two compatible forcing conditions forcing opposite statements. Thus, since weak forcing is expressible in M, we only have to know $A_1, \ldots, A_m$ to completely determine x, i.e., x is constructible from $A_1, \ldots, A_m$. This proof is essentially the same as one we gave in the last section.

LEMMA. If α, β are ordinals and $\bar{\bar{\alpha}} \geq \bar{\bar{\beta}}$ in N, then for some n $\bar{\bar{\alpha}} \cdot \aleph_n \geq \bar{\bar{\beta}}$ in M.

Proof. Let f in N map α onto β. Then f is a set of ordered pairs of ordinals and the previous lemma applies so f really depends only on $A_1, \ldots, A_m$ for some m. Let P_0 force f to be single-valued. Then if $P \supseteq P_0$, and $\gamma < \alpha$, P can weakly force at most one ordinal to be $f(\gamma)$. Since only the part of P which involves $A_1, \ldots, A_m$ is relevant, we see that the range of f is contained in a set whose cardinality in M is at most the cardinality of α multiplied by the number of P which involve only $A_1, \ldots, A_m$ which computed in M is $\aleph_n$ for some n and so the lemma is proved.

COROLLARY. $\aleph_\omega$ in M is uncountable and equals $\aleph_1$ in N.

Proof. $\aleph_\omega$ is uncountable by the lemma and since A_n makes $\aleph_n$ countable, $\aleph_\omega$ is the first uncountable ordinal.

THEOREM. In N, $\aleph_1$ is the supremum of countably many countable ordinals.

Proof. Since $M \subseteq N$, $\aleph_\omega$ in M is still the limit of $\aleph_n$ in M.

By one of our lemmas, every real number x in N is constructible from $A_1,\ldots,A_n$ for some n. Let C_n be the set of such reals. Clearly C_n remains the same if we replace A_i by $\pi_i A_i$. We shall show that C_n is countable in N. Let $x = \bar{c}$, where G_n keeps c fixed. Let R_k for each k be the forcing conditions which involve only $A_1,\ldots,A_n$ and weakly force $k \in c$. It is not hard to see that the sets R_k together determine x if $A_1,\ldots,A_n$ are regarded as fixed. Since the R_k are in M, their cardinality is easily computed in M to be not greater than $\aleph_{n+2}$, which is countable in N, so C_n is countable. Now C_n can also be defined as the set of all real numbers which are constructible from B_1, $\ldots,B_n$. Since we have the set $\{\langle 1,B_1\rangle, \ldots, \langle n,B_n\rangle, \ldots\}$ in N, we can enumerate the sets C_n and we have shown

THEOREM. In N, the continuum is a countable union of countable sets.

Let x be a set of real numbers in N which is well-ordered. Assume G_n keeps x and the well-ordering of x fixed. Then if y is the real number in x which corresponds to the ordinal α, the usual symmetry argument will show that, since α is invariant under G, $m \in x$ (considering $x \subseteq \omega$) is determined by $A_1,\ldots,A_n$. Thus the elements of x are all constructible from $A_1,\ldots,A_n$. We have already seen that this implies that x is a countable set. Hence,

THEOREM. In N, the only well-ordered sets of real numbers are countable.

The process of making cardinals countable can be carried still further. In general, we can repeat the previous process to any cardinal. That is, let β be a cardinal in M, and for each $\alpha < \beta$, let S_α contain a set which makes α countable, together with all finite permutations of this set as we did above. At S_β, we start the usual collecting process and forming new sets by means of formulas without introducing any set analogous to $\{\langle 1,B_1\rangle,\ldots\}$ as before. If $\beta = \aleph_\tau$ where τ is a limit ordinal, we can prove that only countable sets of real numbers can be well-ordered. However, suppose we wish to preserve this fact and still have the countable AC hold. This means, since each S_α with $\alpha < \beta$ is countable, that we must also introduce at S_α a well-ordering of the previous $S_{\alpha'}$, $\alpha' < \alpha$. Thus, on the one hand we are making all $\alpha < \beta$ countable and yet at the same time well-ordering so many sets as to make it very likely that we can piece together the sets that make α countable for $\alpha < \beta$, to show that β itself is countable in N. It turns out that if β is chosen very large, e.g., β inaccessible, that this can be avoided.

11. AVOIDING SM

If one does not care about the construction of actual models, then the independence results of this chapter can be proved in a purely syntactical way without appeal to SM, and indeed entirely within elementary number theory. Of course, one only proves relative consistency results, as explained in Chapter III. One way to proceed is as follows: We completely drop M and consider the forcing notion with no restriction on the ordinal subscripts used. For each statement A, limited or unlimited, one can associate another statement which says P forces A. (Here a statement is regarded as a string of symbols in which ordinals appear.) This is akin to Gödel's association to each statement A, the statement A_L. Now one can prove purely syntactically that if A is a logically valid formula, then all P weakly force A. More precisely, we show how to associate to every proof of A, a proof of the statement that all P weakly force A. Also, the proofs we have given that the axioms hold in N, can now be given purely formally, that every P weakly forces the axioms. Also all the main lemmas about forcing can be proved. One also shows that all P weakly force $\sim$ CH, $\sim V = L$, or whatever else is being considered. Thus, if ZF + AC + $\sim$ CH led to a contradiction, one can show that for some P, P would force a contradiction, which violates one of the basic properties of forcing. Although this point of view may seem like a rather tedious way of avoiding models, it should be mentioned that in our original approach to forcing this syntactical point of view was the dominant point of view, and models were later introduced as they appeared to simplify the exposition. The peculiar role of the countability of M is here entirely avoided.

Another method which probably is easier on first encounter is the following: The axiom SM is only used to construct a countable model. To show that ZF + AC + ~ CH is consistent, it is only necessary to show that every finite set of axioms of ZF + AC + ~ CH is consistent. To prove these finitely many statements in N, requires only finitely many steps and hence uses only finitely many axioms of ZF in M. An analysis of the proof will show explicitly which axioms are used. Now in Chapter II § 8, we proved in ZF that there are models M for any finite set of axioms. Thus, we can work with such an M and avoid axiom SM.

12. GCH IMPLIES AC

We shall now show that GCH implies AC. Since we do not assume AC we must be careful how we define the various concepts related to cardinality. For two sets A and B, we write $\bar{\bar{A}} = \bar{\bar{B}}$ or $\bar{\bar{A}} \leq \bar{\bar{B}}$ if there is a 1-1 map from A onto or into B. We write $\bar{\bar{A}} < \bar{\bar{B}}$ if $\bar{\bar{A}} \leq \bar{\bar{B}}$ and $\bar{\bar{A}} \neq \bar{\bar{B}}$. The Cantor-Bernstein Theorem does not use AC so we know that if $\bar{\bar{A}} \leq \bar{\bar{B}}$ and $\bar{\bar{B}} \leq \bar{\bar{A}}$ then $\bar{\bar{A}} = \bar{\bar{B}}$. We now state GCH as:

If A is an infinite set, there is no B such that $\bar{\bar{A}} < \bar{\bar{B}} < \overline{\overline{2^{A}}}$.

To avoid using exponents we shall write $P_0(A) = A$, $P_n = P(P_{n-1}(A))$ where $P(X)$ denotes the power set of X. Also we drop the use of $=$ so that we write $X = Y$ if $\bar{\bar{X}} = \bar{\bar{Y}}$. The GCH is a rather strong assertion about the existence of various maps since if we ever are given that $A \leq B \leq P(A)$ then there must be a 1-1 map either from B onto A or B onto $P(A)$. Essentially this means that there are so many maps available that we can well-order every set. Let A be an infinite set and we wish to show A can be well-ordered. The proof given here follows that of Sierpinski [24].

LEMMA 1. There is a well-ordered set W, such that $W \subseteq P_4(A)$ and we do not have $W \leq A$.

Proof. Consider all well-orderings of A or of subsets of A. Recall that a well ordering of A is a collection of ordered pairs of A, hence it belongs to $P_3(A)$. (A pair belongs to $P_1(A)$, ordered pair belongs to $P_2(A)$.) Now consider the equivalence relation which considers two well-orderings equivalent if they are order isomorphic. Let W be the set of all the equivalence classes induced by this relation. (The elements of W are thus "ordinal numbers".) The elements of W form a well-ordered set under the usual definition of ordering of well-ordered sets given in Chapter II. If $W \leq A$ then this would mean that W is order isomorphic to one of the well-orderings in W and hence that W is order isomorphic to a proper initial segment of W, which is impossible.

Actually we might expect W to be equal to P(A) in cardinality since if AC were true, W is the set of all ordinals of cardinality equal to $\bar{\bar{A}}$, hence is the next cardinal, which by GCH is the cardinality of P(A). To apply GCH we must construct situations in which W or some set related to W lies between X and P(X) for some X. We first assume that $2P_i(A) = P_i(A)$ for $0 \leq i \leq 4$. Then we know that $P_3(A) \leq W+P_3(A) \leq P_4(A) + P_3(A) = P_4(A)$ by the assumption. Thus by GCH either $W + P_3(A) = P_4(A)$ or $W + P_3(A) = P_3(A)$. To handle the first case we have to prove a simple lemma.

LEMMA 2. If X and Y are sets such that $X + Y = P(2X)$ then $Y \geq P(X)$.

Proof. Here 2X denotes the union of two disjoint copies of X, say X_1 and X_2. If f maps $X \cup Y$ onto $P(X_1 \cup X_2) = P(X_1) \times P(X_2)$, then the image of X, projected onto $P(X_1)$ is not all of $P(X_1)$, since $X_1 < P(X_1)$ and hence if c is not in that projection, f must map some subset of Y onto $c \times P(X_2)$ which means $Y \geq P(X)$.

Now, if $W + P_3(A) = P_4(A) = P(2P_3(A))$, we can conclude by Lemma 2 that $W \geq P_4(A)$, or since $W \leq P_4(A)$, $W = P_4(A)$ and hence $P_4(A)$ is well-ordered, so that since A can be imbedded in $P_4(A)$, A is well-ordered and we are done. If $W + P_3(A) = P_3(A)$, then we have $W \leq P_3(A)$ and we can repeat the same argument to show that either A is well-ordered or $W \leq P_2(A)$. Repeating it again we are reduced to the case

$W \leq P_1(A)$. At this point however, we can exclude the possibility that $W = P_0(A) = A$ by Lemma 1 so that we must conclude $W = P_1(A)$ and that A is well-ordered.

It remains to remove the restriction $2P_i(A) = P_i(A)$ for $0 \leq i \leq 4$. If we put $B = P(A \cup \omega)$ where A is disjoint from ω, one verifies easily that $2P_i(B) = P_i(B)$ for $0 \leq i \leq 4$. For, $2B = P(A + \omega + 1) = P(A+\omega) = B$, since $\omega + 1 = \omega$. Also $B \leq B + 1 \leq 2B$, so $B + 1 = B$. Now $2 \cdot 2^B = 2^{B+1} = 2^B$ and a similar argument shows that $2P_i(B) = P_i(B)$ for all i. Hence the argument applies to B and so B can be well-ordered. Since A can be imbedded naturally in B, A can be well-ordered and our result is proved.

13. CONCLUSION

We are now faced with the problem of how to regard questions such as CH in the light of these independence results. Let us review some possible positions. The finitistic point of view, of course, rejects set theory categorically as referring to non-existent fictions and hence would not feel that CH is either "true" or "false". This attitude would probably be shared by other logicians of a somewhat more liberal persuasion such as Intuitionists. Although it may appear extreme, this viewpoint can be presented quite forcefully and persuasively. In essence, one can feel that set theory is merely a highly successful shell which has nothing to do with "real" sets but at best describes some type of mental process used in describing the real objects such as integers. The great defect with this view is that it leaves unexplained why this fiction is successful and how a presumably incorrect intuition has led us to such a remarkable system. For, as Gödel's Incompleteness Theorem implies, we can generate statements in arithmetic provable in set theory but not in lower systems. To merely reject these statements and forever give up any possibility of deciding them is just as unsatisfactory as not knowing what to do with CH.

Since most mathematicians are more or less idealists in their view that sets actually exist and questions like CH have a meaning, we shall now discuss the problem from this viewpoint. One can feel that our intuition about sets is inexhaustible and that eventually an intuitively clear axiom will be presented which decides CH. One possibility is $V = L$, but this is almost universally rejected. Introducing any of the higher axioms of infinity does not alter the situation since the reader can convince himself that if the model M has a very big cardinal, the introduction of the generic sets used in showing $\sim$ CH consistent, will not affect the size (in no matter what sense the term is used), of the cardinal in N. It seems that only axioms which restrict the character of sets, such as $V = L$, have any hope of proving CH, and these are not very acceptable.

A point of view which the author feels may eventually come to be accepted is that CH is obviously false. The main reason one accepts the Axiom of Infinity is probably that we feel it absurd to think that the process of adding only one set at a time can exhaust the entire universe. Similarly with the higher axioms of infinity. Now $\aleph_1$ is the set of countable ordinals and this is merely a special and the simplest way of generating a higher cardinal. The set C is, in contrast, generated by a totally new and more powerful principle, namely the Power Set Axiom. It is unreasonable to expect that any description of a larger cardinal which attempts to build up that cardinal from ideas deriving from the Replacement Axiom can ever reach C. Thus C is greater than $\aleph_n$, $\aleph_\omega$, $\aleph_\alpha$ where $\alpha = \aleph_\omega$ etc. This point of view regards C as an incredibly rich set given to us by one bold new axiom, which can never be approached by any piecemeal process of construction. Perhaps later generations will see the problem more clearly and express themselves more eloquently.

We close with the observation that the problem of CH is not one which can be avoided by not going up in type to sets of real numbers. A similar undecidable problem can be stated using only the concept of real numbers. Namely, consider the statement that every real number is constructible by a countable ordinal. Instead of speaking of countable ordinals we can speak of suitable subsets of ω. The construction $\alpha \to F_\alpha$ for $\alpha \leq \alpha_0$ where α_0 is countable can be completely described if one merely gives all pairs $\langle \alpha, \beta \rangle$ such that $F_\alpha \in F_\beta$. This in turn can be coded as a real number if one enumerates the ordinals. In this way one only speaks about real numbers and yet has an undecidable statement in ZF.

One cannot push this farther and express any of the set-theoretic questions that we have treated as statements about integers alone. Indeed we can postulate as a rather vague article of faith that any statement in arithmetic is decidable in "normal" set theory, i.e., by some recognizable axiom of infinity. This is of course the case with the undecidable statements of Gödel's theorem which are immediately decidable in higher systems.

[1] G. CANTOR: "Gesammelte Abhandlungen," Berlin, 1932.

[2] P. J. COHEN: "Independence of the Axiom of Choice," Stanford University, 1963.

[3] P. J. COHEN: A Minimal Model for Set Theory, Bull. Amer. Math. Soc. 69 (1963), pp. 537-540.

[4] P. J. COHEN: The Independence of the Continuum Hypothesis, I, II. Proc. Nat. Acad. Sci. U.S.A. 50 (1963), pp. 1143-1148; 51 (1964) pp. 105-110.

[5] P. J. COHEN: "Independence Results in Set Theory," Studies in Logic and the Foundations of Mathematics, North-Holland Publishing Co., Amsterdam, 1965, pp. 39-54.

[6] W. B. EASTON: "Powers of Regular Cardinals," Princeton University Dissertation, 1964.

[7] S. FEFERMAN: Some Applications of the Notions of Forcing Generic, Fund. Math. 55 (1965), pp. 325-345.

[8] S. FEFERMAN and A. LÉVY: Independence Results in Set Theory by Cohen's Method, II. Amer. Math. Soc. Notices 10 (1963), p. 593.

[9] A. A. FRAENKEL and Y. BAR-HILLEL: "Foundations of Set Theory," Amsterdam, 1958, x + 415 pp.

[10] K. GÖDEL: Die Vollständigkeit der Axiome des logischen Functionenkalküls, Monatsh. Math. u. Phys. 37 (1930), pp. 349-360.

[11] K. GÖDEL: Über formal unentscheidbare Sätze der Principia Mathematica und verwandter Systeme, I, Monatsh. Math. u. Phys. 38 (1931), pp. 173-198.

[12] K. GÖDEL: Consistency-proof for the Generalized Continuum Hypothesis, Proc. Nat. Acad. Sci. U.S.A. 25 (1939), pp. 220-224.

[13] K. GÖDEL: What is Cantor's Continuum Problem?, Amer. Math. Monthly 54 (1947), pp. 515-525.

[14] K. GÖDEL: "The Consistency of the Axiom of Choice and of the Generalized Continuum Hypothesis with the Axioms of Set Theory," 4th printing, Princeton, 1958, 69 pp.

[15] S. C. KLEENE: "Introduction to Metamathematics," Amsterdam and Groningen, New York and Toronto, 1952, x + 550 pp.

[16] A. LEVY: Axiom Schemata of Strong Infinity, Pacific J. Math. 10 (1960), pp. 223-238.

[17] A. LÉVY: Independence Results in Set Theory by Cohen's Method. I, II, III, IV. Amer. Math. Soc. Notices 10 (1963), pp. 592-593.

[18] L. LÖWENHEIM: Über Möglichkeiten im Relativkalkül. Math. Ann. 76 (1915), pp. 447-470.

[19] E. MENDELSON: The Axiom of Fundierung and the Axiom of Choice. Arch. Math. Logik Grundlagenforsch 4 (1948), pp. 65-70.

[20] A. MOSTOWSKI: Über die Unabhängigketi des Wohlordnungssatzes vom Ordnungsprinzip. Fund. Math. 32 (1939), pp. 201-252.

[21] J. C. SHEPHERDSON: Inner Models for Set Theory, Part I, J. Symb. Logic 16 (1951), pp. 161-190.

[22] J. C. SHEPHERDSON: Inner Models for Set Theory, Part II, J. Symb. Logic 17 (1952), pp. 225-237.

[23] J. C. SHEPHERDSON: Inner Models for Set Theory, Part III, J. Symb. Logic 18 (1953), pp. 145-167.

[24] W. SIERPINSKI: L'hypothèse généralisée du continu et l'axiome du choix. Fund. Math. 34 (1947), pp. 1-5.

[25] R. SOLOVAY: Independence Results in the Theory of Cardinals, I, II, Preliminary Report, Amer. Math. Soc. Notices 10 (1963), p. 595.

[26] A. TARSKI: "A Decision Method for Elementary Algebra and Geometry," 2nd ed., revised, Berkeley and Los Angeles, 1951, iii + 63 pp.

[27] A. N. WHITEHEAD and B. RUSSELL: "Principia mathematica, 1," 2nd ed., Cambridge, 1925, xlvi + 674 pp.

[28] E. ZERMELO: Untersuchungen über die Grundlagen der Mengenlehre, Math. Ann. 65 (1908), pp. 261-281

A CATALOG OF SELECTED

DOVER BOOKS

IN SCIENCE AND MATHEMATICS

Astronomy

BURNHAM'S CELESTIAL HANDBOOK, Robert Burnham, Jr. Thorough guide to the stars beyond our solar system. Exhaustive treatment. Alphabetical by constellation: Andromeda to Cetus in Vol. 1; Chamaeleon to Orion in Vol. 2; and Pavo to Vulpecula in Vol. 3. Hundreds of illustrations. Index in Vol. 3. 2,000pp. 6⅛ x 9¼.
Vol. I: 0-486-23567-X
Vol. II: 0-486-23568-8
Vol. III: 0-486-23673-0

EXPLORING THE MOON THROUGH BINOCULARS AND SMALL TELESCOPES, Ernest H. Cherrington, Jr. Informative, profusely illustrated guide to locating and identifying craters, rills, seas, mountains, other lunar features. Newly revised and updated with special section of new photos. Over 100 photos and diagrams. 240pp. 8¼ x 11. 0-486-24491-1

THE EXTRATERRESTRIAL LIFE DEBATE, 1750–1900, Michael J. Crowe. First detailed, scholarly study in English of the many ideas that developed from 1750 to 1900 regarding the existence of intelligent extraterrestrial life. Examines ideas of Kant, Herschel, Voltaire, Percival Lowell, many other scientists and thinkers. 16 illustrations. 704pp. 5⅜ x 8½. 0-486-40675-X

THEORIES OF THE WORLD FROM ANTIQUITY TO THE COPERNICAN REVOLUTION, Michael J. Crowe. Newly revised edition of an accessible, enlightening book re-creates the change from an earth-centered to a sun-centered conception of the solar system. 242pp. 5⅜ x 8½. 0-486-41444-2

ARISTARCHUS OF SAMOS: The Ancient Copernicus, Sir Thomas Heath. Heath's history of astronomy ranges from Homer and Hesiod to Aristarchus and includes quotes from numerous thinkers, compilers, and scholasticists from Thales and Anaximander through Pythagoras, Plato, Aristotle, and Heraclides. 34 figures. 448pp. 5⅜ x 8½.
0-486-43886-4

A COMPLETE MANUAL OF AMATEUR ASTRONOMY: TOOLS AND TECHNIQUES FOR ASTRONOMICAL OBSERVATIONS, P. Clay Sherrod with Thomas L. Koed. Concise, highly readable book discusses: selecting, setting up and maintaining a telescope; amateur studies of the sun; lunar topography and occultations; observations of Mars, Jupiter, Saturn, the minor planets and the stars; an introduction to photoelectric photometry; more. 1981 ed. 124 figures. 25 halftones. 37 tables. 335pp. 6½ x 9¼. 0-486-42820-8

AMATEUR ASTRONOMER'S HANDBOOK, J. B. Sidgwick. Timeless, comprehensive coverage of telescopes, mirrors, lenses, mountings, telescope drives, micrometers, spectroscopes, more. 189 illustrations. 576pp. 5⅝ x 8¼. (Available in U.S. only.)
0-486-24034-7

STAR LORE: Myths, Legends, and Facts, William Tyler Olcott. Captivating retellings of the origins and histories of ancient star groups include Pegasus, Ursa Major, Pleiades, signs of the zodiac, and other constellations. "Classic."—Sky & Telescope. 58 illustrations. 544pp. 5⅜ x 8½. 0-486-43581-4

Chemistry

THE SCEPTICAL CHYMIST: THE CLASSIC 1661 TEXT, Robert Boyle. Boyle defines the term "element," asserting that all natural phenomena can be explained by the motion and organization of primary particles. 1911 ed. viii+232pp. $5^3/_8$ x $8^1/_2$.
0-486-42825-7

RADIOACTIVE SUBSTANCES, Marie Curie. Here is the celebrated scientist's doctoral thesis, the prelude to her receipt of the 1903 Nobel Prize. Curie discusses establishing atomic character of radioactivity found in compounds of uranium and thorium; extraction from pitchblende of polonium and radium; isolation of pure radium chloride; determination of atomic weight of radium; plus electric, photographic, luminous, heat, color effects of radioactivity. ii+94pp. $5^3/_8$ x $8^1/_2$. 0-486-42550-9

CHEMICAL MAGIC, Leonard A. Ford. Second Edition, Revised by E. Winston Grundmeier. Over 100 unusual stunts demonstrating cold fire, dust explosions, much more. Text explains scientific principles and stresses safety precautions. 128pp. $5^3/_8$ x $8^1/_2$. 0-486-67628-5

MOLECULAR THEORY OF CAPILLARITY, J. S. Rowlinson and B. Widom. History of surface phenomena offers critical and detailed examination and assessment of modern theories, focusing on statistical mechanics and application of results in mean-field approximation to model systems. 1989 edition. 352pp. $5^3/_8$ x $8^1/_2$. 0-486-42544-4

CHEMICAL AND CATALYTIC REACTION ENGINEERING, James J. Carberry. Designed to offer background for managing chemical reactions, this text examines behavior of chemical reactions and reactors; fluid-fluid and fluid-solid reaction systems; heterogeneous catalysis and catalytic kinetics; more. 1976 edition. 672pp. $6^1/_8$ x $9^1/_4$. 0-486-41736-0 $31.95

ELEMENTS OF CHEMISTRY, Antoine Lavoisier. Monumental classic by founder of modern chemistry in remarkable reprint of rare 1790 Kerr translation. A must for every student of chemistry or the history of science. 539pp. $5^3/_8$ x $8^1/_2$. 0-486-64624-6

MOLECULES AND RADIATION: An Introduction to Modern Molecular Spectroscopy. Second Edition, Jeffrey I. Steinfeld. This unified treatment introduces upper-level undergraduates and graduate students to the concepts and the methods of molecular spectroscopy and applications to quantum electronics, lasers, and related optical phenomena. 1985 edition. 512pp. $5^3/_8$ x $8^1/_2$. 0-486-44152-0

A SHORT HISTORY OF CHEMISTRY, J. R. Partington. Classic exposition explores origins of chemistry, alchemy, early medical chemistry, nature of atmosphere, theory of valency, laws and structure of atomic theory, much more. 428pp. $5^3/_8$ x $8^1/_2$. (Available in U.S. only.) 0-486-65977-1

GENERAL CHEMISTRY, Linus Pauling. Revised 3rd edition of classic first-year text by Nobel laureate. Atomic and molecular structure, quantum mechanics, statistical mechanics, thermodynamics correlated with descriptive chemistry. Problems. 992pp. $5^3/_8$ x $8^1/_2$.
0-486-65622-5

ELECTRON CORRELATION IN MOLECULES, S. Wilson. This text addresses one of theoretical chemistry's central problems. Topics include molecular electronic structure, independent electron models, electron correlation, the linked diagram theorem, and related topics. 1984 edition. 304pp. $5^3/_8$ x $8^1/_2$. 0-486-45879-2

Engineering

DE RE METALLICA, Georgius Agricola. The famous Hoover translation of greatest treatise on technological chemistry, engineering, geology, mining of early modern times (1556). All 289 original woodcuts. 638pp. $6\frac{3}{4}$ x 11. 0-486-60006-8

FUNDAMENTALS OF ASTRODYNAMICS, Roger Bate et al. Modern approach developed by U.S. Air Force Academy. Designed as a first course. Problems, exercises. Numerous illustrations. 455pp. $5\frac{3}{8}$ x $8\frac{1}{2}$. 0-486-60061-0

DYNAMICS OF FLUIDS IN POROUS MEDIA, Jacob Bear. For advanced students of ground water hydrology, soil mechanics and physics, drainage and irrigation engineering and more. 335 illustrations. Exercises, with answers. 784pp. $6\frac{1}{8}$ x $9\frac{1}{4}$. 0-486-65675-6

THEORY OF VISCOELASTICITY (SECOND EDITION), Richard M. Christensen. Complete consistent description of the linear theory of the viscoelastic behavior of materials. Problem-solving techniques discussed. 1982 edition. 29 figures. xiv+364pp. $6\frac{1}{8}$ x $9\frac{1}{4}$. 0-486-42880-X

MECHANICS, J. P. Den Hartog. A classic introductory text or refresher. Hundreds of applications and design problems illuminate fundamentals of trusses, loaded beams and cables, etc. 334 answered problems. 462pp. $5\frac{3}{8}$ x $8\frac{1}{2}$. 0-486-60754-2

MECHANICAL VIBRATIONS, J. P. Den Hartog. Classic textbook offers lucid explanations and illustrative models, applying theories of vibrations to a variety of practical industrial engineering problems. Numerous figures. 233 problems, solutions. Appendix. Index. Preface. 436pp. $5\frac{3}{8}$ x $8\frac{1}{2}$. 0-486-64785-4

STRENGTH OF MATERIALS, J. P. Den Hartog. Full, clear treatment of basic material (tension, torsion, bending, etc.) plus advanced material on engineering methods, applications. 350 answered problems. 323pp. $5\frac{3}{8}$ x $8\frac{1}{2}$. 0-486-60755-0

A HISTORY OF MECHANICS, René Dugas. Monumental study of mechanical principles from antiquity to quantum mechanics. Contributions of ancient Greeks, Galileo, Leonardo, Kepler, Lagrange, many others. 671pp. $5\frac{3}{8}$ x $8\frac{1}{2}$. 0-486-65632-2

STABILITY THEORY AND ITS APPLICATIONS TO STRUCTURAL MECHANICS, Clive L. Dym. Self-contained text focuses on Koiter postbuckling analyses, with mathematical notions of stability of motion. Basing minimum energy principles for static stability upon dynamic concepts of stability of motion, it develops asymptotic buckling and postbuckling analyses from potential energy considerations, with applications to columns, plates, and arches. 1974 ed. 208pp. $5\frac{3}{8}$ x $8\frac{1}{2}$. 0-486-42541-X

BASIC ELECTRICITY, U.S. Bureau of Naval Personnel. Originally a training course; best nontechnical coverage. Topics include batteries, circuits, conductors, AC and DC, inductance and capacitance, generators, motors, transformers, amplifiers, etc. Many questions with answers. 349 illustrations. 1969 edition. 448pp. $6\frac{1}{2}$ x $9\frac{1}{4}$. 0-486-20973-3

Mathematics

FUNCTIONAL ANALYSIS (Second Corrected Edition), George Bachman and Lawrence Narici. Excellent treatment of subject geared toward students with background in linear algebra, advanced calculus, physics and engineering. Text covers introduction to inner-product spaces, normed, metric spaces, and topological spaces; complete orthonormal sets, the Hahn-Banach Theorem and its consequences, and many other related subjects. 1966 ed. 544pp. $6^1/_8$ x $9^1/_4$. 0-486-40251-7

DIFFERENTIAL MANIFOLDS, Antoni A. Kosinski. Introductory text for advanced undergraduates and graduate students presents systematic study of the topological structure of smooth manifolds, starting with elements of theory and concluding with method of surgery. 1993 edition. 288pp. $5^3/_8$ x $8^1/_2$. 0-486-46244-7

VECTOR AND TENSOR ANALYSIS WITH APPLICATIONS, A. I. Borisenko and I. E. Tarapov. Concise introduction. Worked-out problems, solutions, exercises. 257pp. $5^5/_8$ x $8^1/_4$. 0-486-63833-2

AN INTRODUCTION TO ORDINARY DIFFERENTIAL EQUATIONS, Earl A. Coddington. A thorough and systematic first course in elementary differential equations for undergraduates in mathematics and science, with many exercises and problems (with answers). Index. 304pp. $5^3/_8$ x $8^1/_2$. 0-486-65942-9

FOURIER SERIES AND ORTHOGONAL FUNCTIONS, Harry F. Davis. An incisive text combining theory and practical example to introduce Fourier series, orthogonal functions and applications of the Fourier method to boundary-value problems. 570 exercises. Answers and notes. 416pp. $5^3/_8$ x $8^1/_2$. 0-486-65973-9

COMPUTABILITY AND UNSOLVABILITY, Martin Davis. Classic graduate-level introduction to theory of computability, usually referred to as theory of recurrent functions. New preface and appendix. 288pp. $5^3/_8$ x $8^1/_2$. 0-486-61471-9

AN INTRODUCTION TO MATHEMATICAL ANALYSIS, Robert A. Rankin. Dealing chiefly with functions of a single real variable, this text by a distinguished educator introduces limits, continuity, differentiability, integration, convergence of infinite series, double series, and infinite products. 1963 edition. 624pp. $5^3/_8$ x $8^1/_2$. 0-486-46251-X

METHODS OF NUMERICAL INTEGRATION (SECOND EDITION), Philip J. Davis and Philip Rabinowitz. Requiring only a background in calculus, this text covers approximate integration over finite and infinite intervals, error analysis, approximate integration in two or more dimensions, and automatic integration. 1984 edition. 624pp. $5^3/_8$ x $8^1/_2$. 0-486-45339-1

INTRODUCTION TO LINEAR ALGEBRA AND DIFFERENTIAL EQUATIONS, John W. Dettman. Excellent text covers complex numbers, determinants, orthonormal bases, Laplace transforms, much more. Exercises with solutions. Undergraduate level. 416pp. $5^3/_8$ x $8^1/_2$. 0-486-65191-6

RIEMANN'S ZETA FUNCTION, H. M. Edwards. Superb, high-level study of landmark 1859 publication entitled "On the Number of Primes Less Than a Given Magnitude" traces developments in mathematical theory that it inspired. xiv+315pp. $5^3/_8$ x $8^1/_2$. 0-486-41740-9

CALCULUS OF VARIATIONS WITH APPLICATIONS, George M. Ewing. Applications-oriented introduction to variational theory develops insight and promotes understanding of specialized books, research papers. Suitable for advanced undergraduate/graduate students as primary, supplementary text. 352pp. $5^3/_8$ x $8^1/_2$. 0-486-64856-7

MATHEMATICIAN'S DELIGHT, W. W. Sawyer. "Recommended with confidence" by *The Times Literary Supplement,* this lively survey was written by a renowned teacher. It starts with arithmetic and algebra, gradually proceeding to trigonometry and calculus. 1943 edition. 240pp. $5^3/_8$ x $8^1/_2$. 0-486-46240-4

ADVANCED EUCLIDEAN GEOMETRY, Roger A. Johnson. This classic text explores the geometry of the triangle and the circle, concentrating on extensions of Euclidean theory, and examining in detail many relatively recent theorems. 1929 edition. 336pp. $5^3/_8$ x $8^1/_2$. 0-486-46237-4

COUNTEREXAMPLES IN ANALYSIS, Bernard R. Gelbaum and John M. H. Olmsted. These counterexamples deal mostly with the part of analysis known as "real variables." The first half covers the real number system, and the second half encompasses higher dimensions. 1962 edition. xxiv+198pp. $5^3/_8$ x $8^1/_2$. 0-486-42875-3

CATASTROPHE THEORY FOR SCIENTISTS AND ENGINEERS, Robert Gilmore. Advanced-level treatment describes mathematics of theory grounded in the work of Poincaré, R. Thom, other mathematicians. Also important applications to problems in mathematics, physics, chemistry and engineering. 1981 edition. References. 28 tables. 397 black-and-white illustrations. xvii + 666pp. $6^1/_8$ x $9^1/_4$. 0-486-67539-4

COMPLEX VARIABLES: Second Edition, Robert B. Ash and W. P. Novinger. Suitable for advanced undergraduates and graduate students, this newly revised treatment covers Cauchy theorem and its applications, analytic functions, and the prime number theorem. Numerous problems and solutions. 2004 edition. 224pp. $6^1/_2$ x $9^1/_4$. 0-486-46250-1

NUMERICAL METHODS FOR SCIENTISTS AND ENGINEERS, Richard Hamming. Classic text stresses frequency approach in coverage of algorithms, polynomial approximation, Fourier approximation, exponential approximation, other topics. Revised and enlarged 2nd edition. 721pp. $5^3/_8$ x $8^1/_2$. 0-486-65241-6

INTRODUCTION TO NUMERICAL ANALYSIS (2nd Edition), F. B. Hildebrand. Classic, fundamental treatment covers computation, approximation, interpolation, numerical differentiation and integration, other topics. 150 new problems. 669pp. $5^3/_8$ x $8^1/_2$. 0-486-65363-3

MARKOV PROCESSES AND POTENTIAL THEORY, Robert M. Blumental and Ronald K. Getoor. This graduate-level text explores the relationship between Markov processes and potential theory in terms of excessive functions, multiplicative functionals and subprocesses, additive functionals and their potentials, and dual processes. 1968 edition. 320pp. $5^3/_8$ x $8^1/_2$. 0-486-46263-3

ABSTRACT SETS AND FINITE ORDINALS: An Introduction to the Study of Set Theory, G. B. Keene. This text unites logical and philosophical aspects of set theory in a manner intelligible to mathematicians without training in formal logic and to logicians without a mathematical background. 1961 edition. 112pp. $5^3/_8$ x $8^1/_2$. 0-486-46249-8

Physics

OPTICAL RESONANCE AND TWO-LEVEL ATOMS, L. Allen and J. H. Eberly. Clear, comprehensive introduction to basic principles behind all quantum optical resonance phenomena. 53 illustrations. Preface. Index. 256pp. 5³/₈ x 8¹/₂. 0-486-65533-4

QUANTUM THEORY, David Bohm. This advanced undergraduate-level text presents the quantum theory in terms of qualitative and imaginative concepts, followed by specific applications worked out in mathematical detail. Preface. Index. 655pp. 5³/₈ x 8¹/₂. 0-486-65969-0

ATOMIC PHYSICS (8th EDITION), Max Born. Nobel laureate's lucid treatment of kinetic theory of gases, elementary particles, nuclear atom, wave-corpuscles, atomic structure and spectral lines, much more. Over 40 appendices, bibliography. 495pp. 5³/₈ x 8¹/₂. 0-486-65984-4

A SOPHISTICATE'S PRIMER OF RELATIVITY, P. W. Bridgman. Geared toward readers already acquainted with special relativity, this book transcends the view of theory as a working tool to answer natural questions: What is a frame of reference? What is a "law of nature"? What is the role of the "observer"? Extensive treatment, written in terms accessible to those without a scientific background. 1983 ed. xlviii+172pp. 5³/₈ x 8¹/₂. 0-486-42549-5

AN INTRODUCTION TO HAMILTONIAN OPTICS, H. A. Buchdahl. Detailed account of the Hamiltonian treatment of aberration theory in geometrical optics. Many classes of optical systems defined in terms of the symmetries they possess. Problems with detailed solutions. 1970 edition. xv + 360pp. 5³/₈ x 8¹/₂. 0-486-67597-1

PRIMER OF QUANTUM MECHANICS, Marvin Chester. Introductory text examines the classical quantum bead on a track: its state and representations; operator eigenvalues; harmonic oscillator and bound bead in a symmetric force field; and bead in a spherical shell. Other topics include spin, matrices, and the structure of quantum mechanics; the simplest atom; indistinguishable particles; and stationary-state perturbation theory. 1992 ed. xiv+314pp. 6¹/₈ x 9¹/₄. 0-486-42878-8

LECTURES ON QUANTUM MECHANICS, Paul A. M. Dirac. Four concise, brilliant lectures on mathematical methods in quantum mechanics from Nobel Prize-winning quantum pioneer build on idea of visualizing quantum theory through the use of classical mechanics. 96pp. 5³/₈ x 8¹/₂. 0-486-41713-1

THIRTY YEARS THAT SHOOK PHYSICS: THE STORY OF QUANTUM THEORY, George Gamow. Lucid, accessible introduction to influential theory of energy and matter. Careful explanations of Dirac's anti-particles, Bohr's model of the atom, much more. 12 plates. Numerous drawings. 240pp. 5³/₈ x 8¹/₂. 0-486-24895-X

ELECTRONIC STRUCTURE AND THE PROPERTIES OF SOLIDS: THE PHYSICS OF THE CHEMICAL BOND, Walter A. Harrison. Innovative text offers basic understanding of the electronic structure of covalent and ionic solids, simple metals, transition metals and their compounds. Problems. 1980 edition. 582pp. 6¹/₈ x 9¹/₄. 0-486-66021-4